T0350929

Tropical Forest Conservation and Industry Partnership

Conservation Science and Practice Series

Published in association with the Zoological Society of London

Wiley-Blackwell and the Zoological Society of London are proud to present our *Conservation Science and Practice* series. Each book in the series reviews a key issue in conservation today. We are particularly keen to publish books that address the multidisciplinary aspects of conservation, looking at how biological scientists and ecologists are interacting with social scientists to effect long-term, sustainable conservation measures.

Books in the series can be single or multi-authored and proposals should be sent to:
  Ward Cooper, Senior Commissioning Editor, Wiley-Blackwell, John Wiley & Sons,
  9600 Garsington Road, Oxford OX4 2DQ, UK
  Email: ward.cooper@wiley.com

Each book proposal will be assessed by independent academic referees, as well as our Series Editorial Panel. Members of the Panel include:
  Richard Cowling, Nelson Mandela Metropolitan University, Port Elizabeth, South Africa
  John Gittleman, Institute of Ecology, University of Georgia, USA
  Andrew Knight, University of Stellenbosch, South Africa
  Georgina Mace, Imperial College London, Silwood Park, UK
  Daniel Pauly, University of British Columbia, Canada
  Stuart Pimm, Duke University, USA
  Hugh Possingham, University of Queensland, Australia
  Peter Raven, Missouri Botanical Gardens, USA
  Helen Regan, University of California, Riverside, USA
  Alex Rogers, Institute of Zoology, London, UK
  Michael Samways, University of Stellenbosch, South Africa
  Nigel Stork, University of Melbourne, Australia

Previously published

Conservation Science and Practice Series

Tropical Forest Conservation and Industry Partnership: An Experience from the Congo Basin

Edited by

Connie J. Clark and John R. Poulsen

A John Wiley & Sons, Inc., Publication

Library of Congress Cataloging-in-Publication Data

Tropical forest conservation and industry partnership : an experience from the Congo Basin / edited by Connie J. Clark and John R. Poulsen.
　　　p. cm.
　Includes bibliographical references and index.
　ISBN 978-0-470-67373-7 (cloth)
　1. Rain forests–Congo (Brazzaville)–Parc national de Nouabalé-Ndoki–Management.
　2. Buffer zones (Ecosystem management)–Congo (Brazzaville)–Parc national de Nouabalé-Ndoki.　3. Wildlife conservation–Congo (Brazzaville)–Parc national de Nouabalé-Ndoki.　4. Logging–Environmental aspects–Congo (Brazzaville)–Parc national de Nouabalé-Ndoki.　I. Clark, Connie J.　II. Poulsen, John R.
　SD414.C74T76 2012
　577.34096724–dc23

　　　　　　　　　　　　　　　　　　　　　　　　　　　　　　　　2011037508

A catalogue record for this book is available from the British Library.

Typeset in 10.5/12.5 pt Minion by Laserwords Private Limited, Chennai, India
Printed and bound in Malaysia by Vivar Printing Sdn Bhd

1　2012

Contents

Contributors

Thomas Breuer, WCS Congo Program, Wildlife Conservation Society, B.P. 14537, Brazzaville, Republic of Congo.

Terry Brncic, Long-term Ecology Laboratory, Oxford University Centre for the Environment, South Parks Road, Oxford, OX1 3QY, UK.

Connie J. Clark, Nicholas School of the Environment, Duke University, P.O. Box 90328, Durham, NC, USA.

Jean-Claude Dengui, Direction de la Faune et des Aires Protégées, B.P. 98, Brazzaville, Republique du Congo.

Mitchell J. Eaton, NY Cooperative Fish and Wildlife Research Unit, US Geological Survey, Department of Natural Resources, Cornell University, B02 Bruckner Hall, Ithaca, NY 14853, USA.

Paul W. Elkan, WCS Africa Program, International Programs, Wildlife Conservation Society, 2300 Southern Boulevard, Bronx, NY 10460-1099, USA.

Sarah Elkan, WCS Africa Program, International Programs, Wildlife Conservation Society, 2300 Southern Boulevard, Bronx, NY 10460-1099, USA.

Heather E. Eves, 5607 7th Street South, Arlington, VA 22204, USA.

David J. Harris, Royal Botanic Garden Edinburgh, 20A Inverleith Row, Edinburgh, Scotland, EH3 5LR, UK.

Jean Ibara, Ministry of Sustainable Development, Forest Economy and the Environment, Brazzaville, Republic of Congo.

Pierre Kama, Direction de la Faune et des Aires Protégées, B.P. 98, Brazzaville, Republic of Congo.

Kibino Kimbembe, Wildlife Conservation Society, B.P. 14537, Brazzaville, Republic of Congo.

Richard Malonga, Wildlife Conservation Society, B.P. 14537, Brazzaville, Republic of Congo.

Germain A. Mavah, School of Natural Resources and Environment, 3014 Turlington Hall, P.O. Box 117315, University of Florida, Gainesville, FL 32611-7315, USA, and WCS Congo Program, Wildlife Conservation Society, BP 14537, Brazzaville, Republic of Congo.

Olivier Mbani, Wildlife Conservation Society, B.P. 14537 Brazzaville, Republic of Congo.

Miranda Mockrin, Department of Ecology, Evolution, and Environmental Biology, 1200 Amsterdam Avenue, Columbia University, New York, NY 10027, USA, and Rocky Mountain Research Station, USDA Forest Service, 2150 Centre Ave Bldg A, Fort Collins, CO 80521, USA.

Jerome Mokoko-Ikonga, WCS Congo Program, Wildlife Conservation Society, B.P. 14537, Brazzaville, Republic of Congo.

David B. Morgan, Lincoln Park Zoo, Lester E. Fisher Center for the Study and Conservation of Apes, Department of Anthropology, Washington University, One Brookings Drive, Campus Box 1114, Saint Louis, MO 63130, and WCS Congo Program, Wildlife Conservation Society, B.P. 14537, Brazzaville, Republic of Congo.

Antoine Moukassa, Wildlife Conservation Society, BP 14537, Brazzaville, Republic of Congo.

Marcel Ngangoué, Société Likouala Timber, B.P. 14, Bétou, Republic of Congo.

Dominique Nsosso, Direction de la Faune et des Aires Protes, B.P. 98, Brazzaville, Republic of Congo.

Jean Robert Onononag, Goualougo Triangle Ape Project, Wildlife Conservation Society, B.P. 14537, Brazzaville, Republic of Congo.

Jean-Michel Pierre, 7, Rue des Genêts, 66680 Canohès. France.

John R. Poulsen, Nicholas School of the Environment, Duke University, P.O. Box 90328, Durham, NC, USA.

Scott Poynton, The Forest Trust, 4 Chemin des Brumes, 1263 Crassier, Switzerland.

Michael Riddell, School of Geography and the Environment, Oxford University Centre for the Environment (Dyson Perrins Bldg.), University of Oxford, South Parks Road, Oxford, OX1 3QY, UK.

Richard G. Ruggiero, US Fish & Wildlife Service, Division of International Conservation, 4401 N. Fairfax Drive Room 100, Arlington, VA 22203-1622, USA.

Crickette M. Sanz, Department of Anthropology, Washington University, One Brookings Drive, Campus Box 1114, Saint Louis, MO 63130, and WCS Congo Program, Wildlife Conservation Society, B.P. 14537, Brazzaville, Republic of Congo.

Emma J. Stokes, Wildlife Conservation Society, Global Conservation Program, 2300 Southern Boulevard, Bronx, NY 10460-1099, USA.

Samantha Strindberg, Global Conservation Program, Wildlife Conservation Society, 2300 Southern Boulevard, Bronx, NY 10460-1099.

Hannah L. Thomas, Wildlife Conservation Society Congo Program, 2300 Southern Boulevard, Bronx, NY 10460-1099.

Lucas Van der Walt, Olam International Limited, 9 Temasek Blvd., #25-01 Suntec Tower Two, Singapore 038989.

David S. Wilkie, Global Conservation Program, Wildlife Conservation Society, 2300 Southern Boulevard, Bronx, NY 10460-1099.

Foreword

The conservation argument for working in managed landscapes has long been promulgated, and is compelling. In its broadest interpretation conservation is about maintaining the ecological systems that support life on earth, and that requires working across large-scale landscapes. That means seeking for conservation outcomes across the range of different land uses: where nature is protected, where natural resources are being harvested and extracted, where agriculture is the primary human activity and where human impacts are the most intense. That also means working with the full range of institutions, from local government to civil society organizations to multinational corporations.

In a narrower interpretation, conservation is about stewarding biodiversity, from wild species to natural communities. If conservation is to achieve even this narrower goal, it cannot restrict its aspiration to parks and protected areas. Parks and reserves in isolation are too small to protect many species, and ecosystems function at a scale much larger than even the largest protected areas.

Despite the broad consensus that conservationists need to work outside of protected areas and in the broader landscape, few conservationists have however picked up the gauntlet. Fewer still have engaged where it has meant working with extractive industries. This has been a challenge on forested lands in particular, and the typical relationship between conservationists and forestry companies has been one of antagonism. Less than 8% of the world's forests are in protected areas however; if large extents of forests are to be well managed and not converted into other land uses, then conservation organizations need to work with companies on lands that are being logged.

This book explores the successful collaboration between the Wildlife Conservation Society (WCS), an international conservation organization, and the Congolaise Industrielle des Bois (CIB), a European timber company which has been operating timber concessions in the north of the Republic of Congo (Brazzaville) since 1969. The collaboration did not begin especially auspiciously. In the early 1990s, WCS had a management presence in the Nouabalé Ndoki National Park. Surveys in 1989 had identified a forest block of unparalleled biological importance and, in 1993, the Government of Congo had established the park. Immediately to the south and west of the park, CIB had rapidly expanding forestry operations in three timber concessions: Kabo, Pokola and Loundougou. Associated with the forestry operations, the area had experienced dramatic in-migration. In the 1960s, when CIB inherited the small-scale forestry operations of the Société Forestiere du Sangha, the town of Pokola had a population of under 300. By the 1990s, the population had grown to over 11,000 and CIB was employing 1500 workers. Most of the animal protein consumed by the people came from the surrounding forest and there was an active commercial bushmeat trade, with meat being carried on logging trucks leaving the concessions. Hunting, emanating from the towns of Pokola and Kabo and from logging camps, was imperiling the wildlife across the whole landscape and increasingly threatening the park itself.

Difficult negotiations (led by Mike Fay on the WCS side and CIB president Heinrich Stoll) led to a consensus between WCS and CIB in 1995, and the signing of a *Protocole d'Accord* between CIB and leaders of local communities living in the concession areas. That protocol discouraged commercial hunting and established internal regulations prohibiting the transport of bushmeat on logging trucks. Putting operational teeth behind this general agreement was the job of Paul Elkan of WCS and, in 1999, the program was officially launched by WCS, CIB and the Ministry of Forestry. Paul would become the first director of a program that focused on managing and localizing subsistence hunting, controlling the commercial bushmeat trade and providing alternative sources of animal protein. A brigade of eco-guards was recruited under the authority of the Ministry to provide enforcement and WCS assisted with management, fundraising and a program of ecological and law-enforcement monitoring. John Poulsen and Connie Clark, the editors of this volume, began working in the Nouabalé-Ndoki landscape in 2000 and would succeed Paul as WCS program directors in 2005, building on the original foundations.

The collaboration between the conservation organization, the timber company and the government was identified as a test case for conservation in a tropical production forest. The project received early support from the Global Environment Facility (GEF) through the World Bank, USAID's Central African Program for the Environment (CARPE) and the International Timber Trade Organization (ITTO). The project was also viewed with intense suspicion by many conservation and social justice advocates, however. CIB was accused of hiding behind 'green' window-dressing and continuing operations as usual. WCS was accused of being a 'volunteer cheerleader for a billion-dollar industry of exploitation'. The conservation organization was charged with 'selling out' to the exploitative corporation, which itself was only interested in its own profits. Critics expressed concerns that forestry operations would open up the forest, allowing hunting and threatening the livelihoods of forest-dwelling people. Over the years the transaction costs of responding constructively to critics, incurred by all the collaborators, have been significant.

Fifteen years after the initiation of the project, the conservation outcomes are evident although the negative social and environmental consequences of forestry operations in tropical forests need to be recognized. The Nouabalé-Ndoki National Park has been buffered and remains one of the great wildlife refuges in Central Africa. (CIB has sequestered parts of their concessions as strict nature reserves with no logging including, in 2001, the area adjacent to the park known as the 'Goualougo Triangle'.) Within the CIB concessions, overall hunting pressure has been reduced and the commercial bushmeat trade is more controlled. Gorilla, chimpanzee and buffalo are common through the concession, in densities rivaling the park, and elephants still traverse the concession. Additionally, the CIB concessions have continued to strengthen their environmental and social responsiveness. CIB has drawn up an integrated forest management plan and continues to address social issues of forest-dwelling peoples. CIB has worked with Greenpeace and the Tropical Forest Trust, and has responded to recommendations of SGS Qualifor and Rainforest Alliance's Smartwood program to receive Forest Stewardship Council (FSC) certification for its concessions (the first in Africa).

Achieving this success has involved compromises. The collaboration has not been an easy one, as the interests of the partners are not identical and are often antithetical. The collaborators have had to step away from the ideal and engage pragmatically with what is possible. Nevertheless, from a conservation perspective, the net result has been very positive and conservation outcomes are at a scale that could not have been achieved by solely focusing on protected areas.

John G. Robinson
Chief Conservation Officer,
Wildlife Conservation Society

Preface

The idea for this book was conceived on the eve of the ten-year anniversary of a partnership between a government, an international conservation organization and a logging company. These unlikely partners were in the middle of realizing a remarkable achievement: a sustained collaboration to manage wildlife and conserve biodiversity in logging concessions surrounding one of the most pristine national parks in the world, the Nouabalé-Ndoki National Park in the Republic of Congo, Africa. After nearly a decade it had become clear that this new model of conservation, a partnership with the private sector, was making progress and showing discernible outcomes. No project is perfect, but promising results were emerging. The certification of the concessions created one of the largest tracts of tropical forest under management for sustainable timber production. Research showed that despite timber extraction the area still harbored remarkable densities of endangered apes and elephants and, despite challenges, the partners continued to persevere.

The goal of this book is to help to expand the conservation estate by promoting the replication of the partnership in other forests and with different partners. To do this, the book describes and analyzes the history, strategies, activities and management systems of the partnership and its creation, the Buffer Zone Project (BZP). In addition to discussing the principles of conservation and of partnership, and the lessons learned from nearly ten years of experience, the book provides the technical and methodological details to serve as a type of how-to manual. Of course, given the multi-disciplinary nature of biodiversity conservation and management, a real how-to manual would be encyclopedic in scope and encompass biology, economics, anthropology, administration and finance. We hope that this book provides a sufficiently detailed framework for initiating similar projects. The expansion of extractive industries into tropical forest is not limited to the logging sector. The issues addressed, examples given and lessons learned from this initiative cut across industries and are relevant to other extractive industries including mining, oil and agriculture.

The project owes its success to three institutions with the vision to create a partnership and the dedication to keep it going: the Government of Congo (specifically, the Ministry of Forestry Economy or MEF); the Congolaise

Industrielle des Bois (CIB) logging company; and the Wildlife Conservation Society (WCS), an American non-governmental conservation organization.

The project received its funding from the governments of Switzerland, Japan, United States and France through the Global Environment Facility, along with the International Tropical Timber Organization, the United States Agency for International Development's Central African Regional Program for the Environment, United States Fish and Wildlife Service, United States Forest Service, Liz Claiborne Art Ortenberg Foundation, Fonds Français pour l'Environnement Mondial, Columbus Zoo, WCS and CIB.

Multiple people contributed to the project in the field. Notably, Paul Elkan and Sarah Elkan of WCS founded the BZP. With help from their colleagues from MEF and CIB, they built a project from scratch, putting into place the majority of the existing project systems and activities and building the project infrastructure. Paul served as the Project Director from 1998–2003 and then directed the WCS-Congo program as the General Director from 2003–2007. Sarah was the project manager and then the WCS-Congo Financial Director over the same period.

From the WCS side, Mike Fay, John Robinson, Richard Ruggiero, Amy Vedder and Bill Weber all played a role in the development of the partnership. Over the twelve years of the project (1998–2010), Philippe Auzel, Jim Beck, Connie Clark, Mark Dripchack, Antoine Moukassa, Richard Malonga, Germaine Mavah, Suzanne Mondoux, Colby Prevost, John Poulsen and Moise Zoniaba contributed to the management of the project. They were supported by the leadership of WCS-Congo, particularly Bryan Curran, Paul Elkan, Sarah Elkan, Jerome Mokoko-Ikonga and Paul Telfer.

From the CIB side, the management of the company, including Yves Dubois, Jaques Glannaz, Robert Hunink, Jean-Marie Mevellec and Hein-rich Stoll supported the creation and maintenance of the initiative. Fred Glannaz, CIB Forest Management Planner, deserves particular credit for building the relationship between CIB, WCS and MEF and for working with WCS to develop the wildlife management principles at the beginning of the partnership. Later, several people from the forest management team had a hand in continuing the project operations and expanding the program, including Olivier Desmet, Dominique Paget, Jean-Michel Pierre and Lucas Van der Walt.

From the MEF side, Jean-Claude Dengui, Pierre Kama and Jean-Pierre Onday-Otsouma headed the project. Etienne Balenga, Delphin Essieni Elondza, Jean Eyana, Alphonse Ngangambé, Marcel Ngangoué and Pierre

Ngouembé directed the law enforcement unit. Léon Embon served as the MEF envoy to the CIB forestry management unit. Many other MEF agents have served the project and the partnership in various technical capacities, and their contributions are greatly appreciated.

The Buffer Zone Project worked because of the dedication and labors of its Congolese staff. The list of employees is too long to name everyone, but the following people were among those who had a particularly strong impact on the project and its success: Cerylle Assobaum, Gervais Ikeba, Bienvenu Kimbembé, Jean Noel Langa-Langa, Calixte Makoumbou, Richard Malonga, Germain Mavah, Jean Claude Metsampito, Antoine Moukassa, Yves Nganga, Albert Niamazock, Nestor Nianga and Moise Zoniaba. We apologize to the numerous others that we have neglected to mention.

Finally, many local people and communities participated in and contributed to the project, and others have actively engaged in the management of their forest resources.

As ever in an enterprise such as this, a large number of people have contributed to bringing this book to fruition. We are grateful to those who have contributed to chapters through data, writings and boxes, and for their good humor throughout. We also thank those who took the time to review chapters: some of the authors themselves, as well as Olivier Desmet, Paul Elkan, Liz Forest, Suzanne Mondoux, Dominque Paget, Jack Putz, Paul Telfer, John Waugh and David Wilkie.

Finally, it is our hope that this book will highlight the challenge of biodiversity conservation in a rapidly changing world where the human appetite for resources is always in potential conflict with conservation. We proffer one practical solution for mitigating the impacts of natural resource extraction, to bring us a step closer to securing the future of tropical forests and all their diversity.

List of Acronyms

BACIPS	Before-after-paired-control impact series (BACIPS)
BZP	Buffer Zone Project
CARPE	Central Africa Regional Program for the Environment (USAID)
CBD	United Nations Convention on Biological Diversity
CBFP	Congo Basin Forest Partnership
CBM	Community-based management
CFA	Central African franc (Coopération financière en Afrique centrale)
CM	Collaborative management
CIB	Congolaise Industrielle des Bois
DLH	Dalhoff Larsen and Horneman group
FMU	Forest management unit
FSC	Forest Stewardship Council
GIS	Geographic information system
GPS	Global positioning system
GOC	Government of Congo
INCEF	International Conservation and Education Fund
ITTO	International Tropical Timber Organization
IUCN	International Union for Conservation of Nature
LCAOF	Liz Claireborne Art Ortenberg Foundation
LEM	Law enforcement monitoring
LTCR	Lac Télé Community Reserve
MDDEFE	Ministère de le Développement Durable, Economie Forestière, et de l'Environnement
MEF	Ministère de l'Economie Forestière
MOU	Memorandum of understanding
NGO	Non-governmental organization
NNNP	Nouabalé-Ndoki National Park
NP	National Park
NTFP	Non-timber forest product
PROGEPP	Projet de la Gestion des Ecosystèmes Périphériques au Parc

PSPC	Private sector partnership for conservation
PTA	Principal technical advisor
REDD	Reducing Emissions from Deforestation and forest Degradation
RIL	Reduced-impact logging
ROC	Republic of Congo
SNBS	Société Nouvelle des Bois de la Sangha
SPTD	Semi-permanent transect design
STN	Sangha Tri-national Network
TFT	The Forest Trust
UNFCCC	United Nations Framework Convention on Climate Change
USAID	United States Agency for International Development
USLAB	Unité de Surveillance de Lutte Anti-Braconnage
USFWS	United States Fish and Wildlife Services
WCS	Wildlife Conservation Society

Introduction

Connie J. Clark and John R. Poulsen

Nicholas School of the Environment, Duke University, Durham, NC

The end of an era

Tropical rainforests harbor roughly 50% of the world's biotic diversity and provide globally critical ecosystem functions, including 40–50% of the global net primary productivity of terrestrial vegetation (Malhi & Grace, 2000; Houghton, 2008). They also provide valuable products and services to support the livelihoods of rural people. More than 1.6 billion people depend on forests for their livelihoods (Eba'a Atvi *et al.*, 2009). Forest cover continues to decline at a rate of 11.5 million hectares per year however, an area larger than Iceland or Liberia (Hansen *et al.*, 2010). Deforestation and forest degradation are causing an unprecedented loss of biodiversity and ecosystem function (Chapin *et al.*, 2000; Thomas *et al.*, 2004) and are contributing 15–20% of the atmospheric CO_2 that is the primary driver of global climate change (IPCC, 2007).

For the last century, forest conservation has focused largely on the establishment of protected areas where resource extraction is mostly prohibited. Although 113 million hectares has been protected, this amounts to only 8.6% of remaining tropical forests[1] (Nelson & Chomitz, 2009). By itself, the existing system of protected areas is inadequate to prevent continued loss of biodiversity or to protect all ecosystem services (Pimm & Lawton, 1998; Soule & Sanjayan, 1998; Fagan *et al.*, 2006).

[1] This estimation edges over, 20% when multi-use management zones, such as those described in this book, are included in protected area calculations.

Tropical Forest Conservation and Industry Partnership: An Experience from the Congo Basin, First Edition. Edited by Connie J. Clark and John R. Poulsen.
© 2012 Wildlife Conservation Society. Published 2012 by John Wiley & Sons, Ltd.

Outside of protected areas, the image of tropical forests as endless expanses of remote wilderness is, for the most part, an illusion. With few exceptions, tropical forests today are heterogeneous landscapes bisected by roads, dotted with villages and towns, and exploited by smallholders and big businesses for ranching, agriculture, mining, oil and logging (Figure 1.1). Most tropical and high biodiversity forests lie in developing nations whose governments are either unable or unwilling to manage the forest estate sustainably, particularly when large-scale extractive industries generate significant revenue

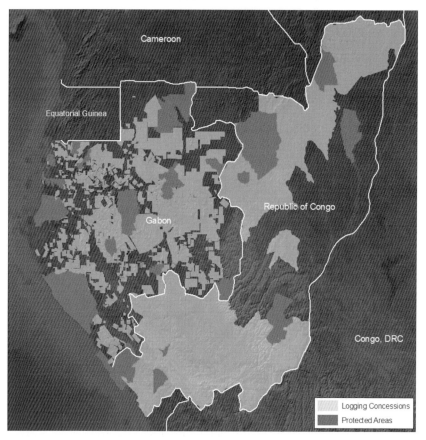

Figure 1.1 **Map of the Republic of Congo and Gabon with Protected Areas represented in dark gray and logging concessions represented in light gray. Data provided by WRI. Map created by Greg Fiske, WHRC.**

(Barrett *et al.*, 2001; Milner-Gulland *et al.*, 2003; Smith *et al.*, 2003). When poor nations must rely on forest exploitation as a source of revenue to build their economies, the possibility of setting aside protected areas vanishes[2]. We must therefore extend conservation efforts beyond protected areas to include places where economic considerations prevail. An obvious place to start is that part of the forest estate dedicated to timber extraction.

Extending the conservation estate to production forests

In addition to protected areas, community forests and production forests are major land uses in tropical forests. Companies leasing logging concessions control significant areas of tropical forest. Managing production forests for biodiversity and ecosystem services could vastly increase the conservation estate. Worldwide, nearly 30% (350 million hectares) of natural tropical forests are destined for logging. The demand for timber is expected to increase from 1.6 billion m^3 to 1.9 billion m^3 during 2010–2015 (Kirilenko & Sedjo, 2007). Demand for fuel wood is also expected to greatly increase pressure on forests in coming decades (Raunikar *et al.*, 2009). With timber plantations occupying only 3% of forested lands, demand for wood in the short term will entail opening up frontier forests (Laporte *et al.*, 2007) or intensifying harvest in already logged forests (FAO, 2009).

The history of logging is one of poorly organized operations and destructive harvests that leave behind a sea of residual damage, rendering forests susceptible to drought, fire and conversion to other uses (Holdsworth & Uhl, 1997; Nepstad *et al.*, 1999; Laurance, 2000; Putz *et al.*, 2000; Asner *et al.*, 2006). The extraction of large, commercially valuable trees creates gaps and modifies the structure of forests for decades to come. Timber harvest can cause widespread collateral damage to remaining trees, subcanopy vegetation, soils, and hydrological processes, resulting in increased erosion and fire, and decreased carbon storage, flora and fauna (Holdsworth & Uhl, 1997; Nepstad *et al.*, 1999; Laurance, 2000; Putz *et al.*, 2000; Asner *et al.*, 2006). Logging operations also

---

[2] Within the UN Framework Convention on Climate Change (UNFCCC), negotiators are designing a mechanism for compensating developing (largely tropical) nations that succeed in reducing carbon emissions from deforestation and forest degradation, known by the acronym from reducing emissions from deforestation and degradation (REDD) (Gullison *et al.*, 2007). This program may incentivize governments to conserve forest by providing revenues to keep forests standing.

open roads into once-remote forests (Figure 1.2). This allows hunters and slash-and-burn farmers to penetrate into forests, resulting in defaunation and further deforestation (Robinson *et al.*, 1999; Wilkie *et al.*, 2000; Asner *et al.*, 2006; Peres *et al.*, 2006; Poulsen *et al.*, 2009).

Many of the impacts of logging are obligate attributes of the industry (e.g., tree extraction, road and skid trail construction; Box 1.1). While they cannot be avoided, their impact can be reduced (e.g., Putz *et al.*, 2008). Indeed, a growing number of companies are committed to adopting reduced-impact logging (RIL) techniques. RIL consists of technologies and practices that are designed to minimize the environmental impacts associated with industrial timber harvesting operations. These include, among other things, pre-harvest inventories and mapping of individual trees, the planning of roads, skid trails and landings and the use of controlled felling and bucking techniques. RIL is an essential component of sustainable timber harvesting prescriptions where operations occur on slopes of less than $40°$ (Putz *et al.*, 2001; Sist & Ferreira, 2007). RIL can be both profitable and renewable, while maintaining much of the carbon stock and biodiversity (Asner *et al.*, 2006; Azevedo-Ramos *et al.*, 2006; Putz *et al.*, 2000, 2008; Clark *et al.*, 2009).

Box 1.1 **Direct and indirect threats of logging**

The threats of anthropogenic activities on the forest and on biodiversity are both direct and indirect. Direct threats are the result of the activity and include: (1) fragmentation of the forest by logging roads; (2) opening of canopy through timber extraction; (3) depletion of timber species; (4) removal of aboveground biomass; and (5) erosion from roads and skid trails.

Indirect threats are often unintended consequences of an activity (e.g., increase in hunting levels because logging roads open access to frontier forest). The indirect threats of logging are somewhat more diverse in their impact. They can be ecological: (1) changes in abundances of animals; (2) loss of biodiversity; and (3) changes in ecological functions (e.g., seed dispersal, regeneration dynamics, etc.). They can also include socio-economic impacts: (1) migration and population growth; (2) increase in levels of hunting and poaching; (3) development of markets for natural resources; (4) encroachment of agriculture; and (5) impacts on human health. The indirect threats that drive overhunting in the tropics are not unique to logging. Without active management, large-scale operations of other extractive industries such as mining, oil exploitation, and industrial agriculture can lead to overhunting, local extinction of some species and the loss of ecosystem services.

These direct and indirect threats combine to contribute to two linked phenomena: the harvest of bushmeat and the loss of biodiversity. The 'bushmeat crisis' refers to the enormous impact that commercial hunting for the meat of wild animals is having on wildlife populations around the world. Commercial hunting has become the most significant immediate threat to the future of wildlife. This threat to wildlife is a crisis because it is rapidly expanding to countries and species which were previously not at risk. The bushmeat crisis is also a human tragedy: the loss of wildlife threatens the livelihoods and food security of rural populations dependent on wildlife as a staple or supplement to their diet. The related 'biodiversity crisis' refers to the rapid loss of species that make up biodiversity – the full complement of species that inhabit an area. Many biologists believe Earth is entering a sixth mass extinction event, one that is the direct consequence of human activities including over-exploitation, habitat destruction and introduction of alien invasive species and pathogens.

Ironically, some of the most devastating impacts of the logging industry on tropical forests are non-obligate and could be avoided with appropriate management practices, including the implementation of mitigation strategies. These secondary, non-obligate impacts of logging include: the unsustainable harvest of non-timber forest resources and, in particular, wildlife; the expansion of destructive slash-and-burn agriculture that results from access to previously inaccessible lands; inequitable distribution of wealth; and poor placement of roads, logging camps and sawmills (e.g., building them near ecologically sensitive areas). Mitigating these non-obligate impacts of logging are traditionally outside of the perceived responsibilities of logging companies, because they are not directly related to logging techniques and operations. Moreover, the conservation community has not pressed for mitigation.

Extending conservation efforts to production forests provides an opportunity to reduce numerous threats to tropical forests, biodiversity and ecosystem services (Box 1.2). In theory, given the large size and varied habitats of logging concessions, logging operations that mitigate both their obligate (direct) and non-obligate (indirect) impacts could help to protect and preserve forest integrity and biodiversity across a landscape. This strategy would contribute in particular to the conservation of wide-ranging species that cannot be contained within the borders of protected areas (Sanderson *et al.*, 2002; Elkan *et al.*, 2006; Blake *et al.*, 2007, 2008; Clark *et al.*, 2009; Stokes *et al.*, 2010).

Box 1.2 Can forestry be conducive to biodiversity conservation?

A number of initiatives relating to improved forest management address biodiversity issues. In recent years, both the International Tropical Timber Organization (ITTO) and International Union for Conservation of Nature (IUCN) emphasized the importance of biodiversity by calling for: (1) an enhanced role for tropical production forests as components of multi-functional landscapes that contribute to biodiversity conservation at different spatial scales; (2) improved understanding of the impacts of forest management on biodiversity; (3) adaptation of forest management practices at all spatial scales to favor the conservation of native biodiversity; (4) improved ecological processes in tropical production forests provided by the presence of locally adapted native biodiversity; and (5) improved practical forest management at all spatial scales aimed at retaining native biodiversity (IUCN, 2006).

Perhaps the best model of forest management that explicitly addresses biodiversity conservation and the preservation of ecosystem functions is that of the Forest Stewardship Council (FSC). The FSC's set of Principles and Criteria for forest management includes ten principles illustrated by a number of criteria, several of which directly address the indirect impacts of logging on biodiversity and ecosystem function. For example, Principle 6, Criterion 2 states: "Safeguards shall exist which protect rare, threatened and endangered species and their habitats (e.g., nesting and feeding areas). Conservation zones and protection areas shall be established appropriate to the scale and intensity of forest management and the uniqueness of the affected resources. Inappropriate hunting, trapping and collecting shall be controlled".

While these initiatives emphasize the importance of biodiversity conservation and protection of ecosystem functions, they have been criticized for failing to specify standards or actions to mitigate biodiversity loss in production forests (Bennett, 2001; Poulsen *et al.*, 2009). Despite the development of principles and criteria that encourage the conservation of biodiversity and ecosystem processes in production forests, very few examples of well-managed forestry concessions exist in the Congo Basin.

A new paradigm: Private-sector partnerships for conservation

Extending conservation activities into production forests requires innovative strategies that look beyond a strict protection paradigm toward a conservation model that accommodates the resource needs and land-use practices of multiple stakeholders. These conservation strategies will need to integrate national governments, industry, conservation NGOs and local communities in partnerships that promote environmentally appropriate forest use and biodiversity conservation across multi-use landscapes.

It is our hope that this book will facilitate the development of partnerships to mitigate the non-obligate impacts of logging on biodiversity, ecosystem services and livelihoods of local populations, with particular emphasis on wildlife management. The processes, systems and methodologies presented in this book were developed and tested in forestry concessions in northern Republic of Congo, where the Congolaise Industrielle des Bois (CIB) logging company has concessionary rights over 1.3 million hectares of tropical forests adjacent to a protected area network (Figure 1.3). In 1999, the Wildlife Conservation Society (WCS), the Congo's Ministry of Forest Economy (MEF)[3] and the CIB formed a novel alliance to reduce the negative impacts of logging on wildlife and biodiversity in timber concessions adjacent to the Nouabalé-Ndoki

[3] At the project inception, the ministry was named the Ministry of Forest Economy but in 2009 changed to the Ministry of Sustainable Development, Forest Economy and the Environment (MDDEFE). We use the more succinct acronym (MEF) throughout the book.

Figure 1.2   Pokola village from the air. Pokola is the headquarters for the Congolaise Industrielle des Bois (CIB) and an example of logging towns built in frontier forest. Photo by Kent Redford.

National Park (Figure 1.3). The partnership was formally called the *Projet de la Gestion des Ecosystèmes Périphériques au Parc* (PROGEPP), which we have translated and shortened to the Buffer Zone Project (BZP). Throughout the development of the BZP, the partners sought to achieve seemingly contradictory goals: to conserve biodiversity and improve the livelihoods of rural people while promoting economic development though timber exploitation.

The BZP partnership is an example of an emerging conservation model, which we refer to as a private-sector partnership for conservation (PSPC), that extends the conservation estate while integrating the interests and needs of multiple stakeholders. There have been few previous attempts by non-governmental organizations (NGOs), governments, logging companies and communities to find common ground and work together for conservation (see Mayers & Vermeulen, 2002 for similar examples of community-company partnerships). Thus, very little information is available to guide the creation and implementation of PSPCs. The goal of this book is to fill this need using the BZP as a case study of how such partnerships might be structured and implemented. Although we subscribe to the conservation model developed by the BZP, we hold no misconception that it has been implemented perfectly or that the approach can fit all situations and solve all problems. It is our hope that

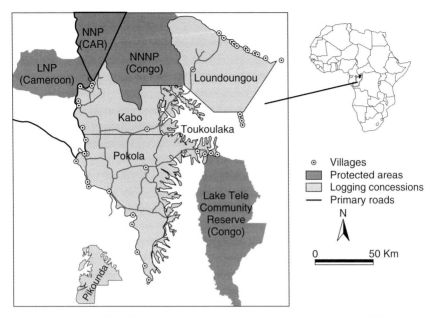

Figure 1.3   **Map of the five logging concessions (Kabo, Pokola, Toukoulaka, Loun-doungou and Pikounda; 1.3 million hectares) in northern Republic of Congo. The concessions lie adjacent to the Lake Télé Community Reserve (440,000 hectares) and the Sangha Tri-national Network (760,000 hectares) of protected areas, including the Nouabalé-Ndoki National Park (NNNP) in Republic of Congo, the Dzanga-Ndoki National Park (NNP) in Central African Republic and the Lobéké National Park (LNP) in Cameroon. Thick black lines indicate the borders between the three countries. Protected areas are shaded in dark gray, CIB concessions are shaded in light gray and unshaded areas are logging concessions operated by other companies or forest. Some roads outside the concessions are not shown.**

the lessons learned from the BZP can help in the design and implementation of similar initiatives at other sites and with other industries, and that the BZP can be used as a demonstration platform for such new initiatives.

We wish to underscore that the effective conservation of tropical forests is not a task that can be tackled by conservation organizations alone. The inclusion of stakeholders in co-management projects may prove to be key to decreasing deforestation and forest degradation and the associated loss of biodiversity and ecosystem functioning. We hope this book will be of interest and use to any business, community group, non-governmental organization, government agency or individual seeking to conserve and manage tropical forests.

Case study: Buffer Zone Project

The social, economic and historical context of any area under conservation determines, to some degree, its challenges, successes and failures. We therefore begin with an overview of the context in which the BZP was developed.

An overview of tropical forestry and conservation in the Congo Basin

The Congo Basin holds the world's largest area of tropical forest after the Amazon Basin. Whereas rainforests in West and East Africa have been reduced to 8–12% of their former extent, Congo Basin forests still encompass nearly 60% of their original distribution and include large tracts of contiguous forest (Naughton-Treves & Weber, 2001; Wilkie & Laporte, 2001). Even so, these forests are increasingly threatened by the expansion of extractive industries. Selective logging is the most extensive industry in the region: timber concessions occupy 30–45% of all remaining tropical forests and over 70% of forests in some countries (Global Forest Watch, 2002; Laporte et al., 2007).

Forestry contributes to the national economy of Congo Basin countries by providing revenue and jobs (Table 1.1). Logging accounts for between 0.22% (Equatorial Guinea) to 6.3% (Central African Republic) of gross domestic

Table 1.1 **Economic importance of logging in the Congo Basin. The contribution of the forestry sector to employment and gross domestic product in 2006 (FAO, 2009).**

Country	Employment		Gross value added	
	1000	% Labor force	US $ million	% GDP
Cameroon	20	0.3	324	1.9
CAR	4	0.2	144	11.1
Congo	7	0.5	72	1.1
DRC	6	0	186	2.3
Equatorial Guinea	1	0.5	87	0.9
Gabon	12	1.9	290	3

product (Eba'a *et al.*, 2009). It is second only to mining/oil extraction in the creation of employment (Congo Basin Forest Partnership, 2006; Congolaise Industrielle des Bois, 2006). By comparison, the oil industry accounts for up to 52% of GDP in some countries (Republic of Congo; http://congo-brazzaville.org/economy). Some experts estimate that depletion of oil stocks will occur within a decade, so tropical forestry is expected to grow in importance to national economies (World Bank Group, 2010). Although positive for economic development in the short term, logging is a challenge to conservation not only because of its direct impacts on the environment but also because it serves as a catalyst to other activities that severely threaten ecosystem services and the biological integrity of forest systems (Box 1.3).

Box 1.3 **Logging, roads and bushmeat hunting (David S. Wilkie, Wildlife Conservation Society)**

Looking at a Global Forest Watch map brings home the extent of Central Africa's forest estate allocated to timber companies (e.g., Figure 1.1). If loggers in Congo and Gabon were to clearcut their concessions, over 70% of the nation's land area would be deforested in 30 years. Thankfully (from a conservation perspective), timber companies active in this region are only interested in felling a few high-value species such as African mahogany (*Entandrophragma* spp., Meliaceae family) and Ocoumé, an excellent source for plywood (*Aucoumea klaineana*, Bignoniaceae family); rarely do they harvest more than 1–2 large trees per hectare.

In fact, high-resolution satellite images taken before and after logging in the northern Republic of Congo show almost no change in the forest (with the exception of a spider-web of roads). After logging, Central Africa's forests appear to be mostly intact; they are not however, because satellite images fail to show the true impacts. It is the roads that are the problem. Although necessary for getting the logs from the forest to regional and international markets, roads and the logging vehicles that traverse them provide easy access to pristine, frontier forests stocked with the highest wildlife biomass of any tropical forest.

Before the loggers arrived, local people throughout the forests of Central Africa were seldom connected to markets and therefore had no history of growing commercial crops. When the timber companies arrive, they build roads; the steady stream of logging trucks traveling from the forest to urban centers and ports offer families their first opportunity to access markets. High-value export crops such as coffee or cacao take years to grow and low-value crops such as cassava and plantains are unprofitable as producers closer to markets have lower transportation costs and can sell for less. Bushmeat, on the other hand, is readily available and of high value. When loggers arrive, forest people find commercial bushmeat hunting to be the best short-term option for taking advantage of the situation. The problem is that such commercial hunting can strip a forest of its wildlife in a matter of months.

While logged forests may look relatively intact from a satellite orbiting overhead, they will be empty of wildlife if commercial hunting has not been controlled. This loss of wildlife is a disaster in the long-term for local communities. It may also be a disaster for the forests because it can change tree species composition and distribution. Large-bodied wildlife such as elephants, gorillas, chimpanzees, buffalo and pigs are no longer around to consume, spread and facilitate germination of the seeds of many tree species.

Consequently, the removal of logs is not the most important threat associated with the vast area of Central Africa's forests under forest concessions. The most important threat is uncontrolled hunting of wildlife for the bushmeat trade. Unless loggers learn how to manage wildlife in their concessions, and are provided the regulatory and economic incentives to do so, large-bodied wildlife may disappear from most of Central Africa's forests in the next 30 years.

Conservation efforts in the Congo Basin are largely focused on a network of 188 protected areas, comprised mostly of national parks[4] (Eba'a Atvi et al., 2009). These account for a little over 9% of the national territories of the six forest-rich nations: Cameroon, the Republic of Congo, the Democratic Republic of Congo, the Central African Republic, Gabon and Equatorial Guinea.

---

[4] This figure does not include recreational hunting zones, hunting reserves and community reserves. With these, there are 341 protected areas representing 14% of the area of Congo Basin countries.

Northern Republic of Congo

The Republic of Congo's population is small (approximately 4 million people in 2010, or approximately 11.7 inhabitants per square kilometer), but it is increasing at an annual rate of 2.7% (https://www.cia.gov/library/publications/the-world-factbook/geos/cf.html, accessed October 18, 2010). Sixty-one percent of the population is concentrated in the urban centers of the capital Brazzaville and the primary industrial port of Pointe Noire; rural areas remain sparsely populated. For example, 648,000 people inhabit northern Congo at an average density of only 2 inhabitants per square kilometer. Low population density coupled with limited agricultural development in forested areas has historically kept both land conversion and biodiversity loss to low levels relative to other forested tropical countries.

Over the past 20 years logging has opened up Congo's forests, allowing hunters to penetrate deep into previously remote areas (Figure 1.4). The meat of wild animals, killed for subsistence or commercial purposes throughout the humid tropics, is commonly referred to as bushmeat. With logging, the scale of the bushmeat trade has increased dramatically (Box 1.3). Because wild game provides local and indigenous populations with a substantial portion of their protein and economic resources (Chapter 6), uncontrolled hunting threatens game species along with biodiversity. Local extirpation of game species would

Figure 1.4   **A village man heading out to hunt with his shotgun. Hunters gain access to once remote forests by roads opened for logging. Photo by Michael Riddell.**

leave some people with few alternative sources of protein, turning wildlife conservation into a food security issue.

Protected areas are the core of Congo's policy for ensuring long-term protection of natural resources and biodiversity. In total, Congo's 15 protected areas cover 3.7 million hectares and represent 10.8% of the total surface area. In the northern part of the country, protected areas cover over 2 million hectares and include the Odzala-Koukoua National Park (OKNP), Nouabalé-Ndoki National Park (NNNP) and the Lac Télé Community Reserve (LTCR) (Figure 1.1). These protected areas are relatively intact and are rich in plant and animal diversity. Much of Congo's wildlife occurs outside of protected areas, however, and most of it is within the logging concessions that cover some 79% of the forest domain (Global Forest Watch, 2002). Properly managed production forests could therefore offer important opportunities for increasing the scope of forest conservation activities in Congo. Apparently recognizing this opportunity, the government has passed a series of laws that, if implemented, would served as a springboard for the extension of conservation activities within production forests (Box 1.4). The CIB forestry company was the first to meet the criteria for well-managed production forest as outlined in Congolese law; others are following their lead.

Box 1.4 Management principles in the Congolese Forest Law pertaining to biodiversity conservation and the rights of local people

1. *Land-use planning*. National Forestry Management Directives include five types of land-use categories: (1) production zones set aside for logging operations and economic production; (2) conservation zones established to conserve biodiversity, wildlife and landscapes; (3) protection zones to safeguard fragile habitats such as watersheds and watercourses; (4) community development zones reserved for use by local populations to exploit natural resources for their livelihoods and community development; and (5) dedicated research zones.
2. *Multi-resource inventories*. Inventories are required to plan extraction cycles as a function of the estimated regenerative capacity of the entire forest resource base. The main focus of resource inventories is

valuable timber species, but surveys of wildlife and other non-timber forest products should also be included.

3. *Participatory management with local communities.* Legitimate interests and rights of local communities must be incorporated into the management planning process. However, national directives do not detail how the policy on participatory management should be implemented. Socio-economic surveys are needed to understand traditional and current patterns of land use by local communities.

4. *Enforcement of hunting laws.* The 2002 Congolese forestry law requires logging companies to pay for a law enforcement unit (*Unité de Surveillance de Lutte Anti-Braconnage* or USLAB) to enforce hunting and wildlife laws within their concessions.

Nouabalé-Ndoki National Park and the Buffer Zone Project

The Nouabalé-Ndoki National Park represents one of the last 'untouched' protected areas in Central Africa. The park has never been commercially logged, although hunter-gatherer populations have inhabited the region for approximately 40,000 years. Evidence of iron smelting and burning sites has been found dating as early as 1240 BP (Lanfranchi *et al.*, 1998; Zangato, 1999; Brncic *et al.*, 2009)[5]. Although originally designated as a logging concession, the park was officially created in 1993 and has since been actively managed through a partnership of the Ministry of Forest Economy and the Wildlife Conservation Society. It preserves 426,000 hectares of tropical forest and contains a diversity of habitats, including a large block of semi-deciduous forest, swamp forest and mono-dominant *Gilbertiodendron* stands believed to have been unaltered by humans for centuries. Many species of globally endangered mammals whose numbers have been drastically reduced at other sites across the Congo Basin by habitat loss and hunting inhabit these forests, including forest elephants, lowland gorillas, chimpanzees, bongo, buffalo, leopard and giant forest hogs (Laurance *et al.*, 2006; Blake *et al.*, 2007). The park also harbors many common species including six species of duiker,

---

[5] Note that the last human settlements in the park date back to 900 years ago, and there is no evidence that people inhabited the park in the last century (Curran *et al.*, 2009; Eves & Ruggiero, 2000).

eight species of diurnal monkeys, sitatunga and the red river hog. More than 300 bird species and 1000 plant species have been recorded within the borders of the park, and the reptiles, amphibians and insects have yet to be comprehensively surveyed.

The Nouabalé-Ndoki National Park forms part of the Sangha Tri-National Network of protected areas. To its west and northwest are the neighboring Lobéké and Ndoki national parks (Figure 1.3). On the Congo side, logging concessions surround the Nouabalé-Ndoki National Park making it the last block of undisturbed forest in the northern part of the country. Selective logging has occurred in the area since the 1960s. Operations were relatively limited until the end of civil war in 1999 which resulted in a rapid expansion of the sector, opening of concessions and construction of roads by large commercial operators. The Kabo (300,000 hectares) and Pokola concessions (560,000 hectares) to the south have been the sites of commercial timber exploitation since the 1970s. Loundoungou (386,000 hectares) to the east and Mokabi (375,000 hectares) to the north were attributed to international timber companies in 2001. The increase in logging activity in northern Congo in the late 1990s and early 2000s opened the park to new threats.

The Buffer Zone Project was created to combat the direct and indirect threats of logging to this area (Box 1.5). The project helped protect the park by managing wildlife, conserving biodiversity and reducing the impact of logging to the forest in the logging concessions near and around the park (Kabo, Pokola, Loundoungou[6] and Pikounda). These concessions formed a buffer zone around the park that also contributed to the conservation of the Sangha Tri-national Network (Figure 1.3).

Box 1.5 **Pre-project situation (Richard G. Ruggiero, US Fish and Wildlife Service)**

In the early 1990s, life on the Sangha River continued as it had throughout much of the 20th century. Local residents still relied on forests that were increasingly penetrated by foreign logging companies and the flood of workers, hunters and traders. Logging camps became

---

[6] In 2009 the Toukoulaka and Loundoungou concessions were merged into a single concession and renamed 'Loundoungou'.

towns with thousands of workers and their families (Figure 1.2). The cash economy produced lucrative markets for bushmeat. Entrepreneurs (both locals and outsiders) who could afford firearms, wire snares and transportation took advantage of open access to harvest wildlife to sell locally at the logging camps and in 'export markets' throughout Congo and beyond.

The history of wildlife exploitation for export is rooted in the colonial concessionary system (Hardin, 2011). The recent advent of large-scale industrial logging and the concomitant roads, infrastructure and human populations has resulted in unsustainable levels of hunting. The ease of hunting due to modern technology, accessibility using logging roads and the market demand in distant cities quickly demonstrate the disparity between the demand for bushmeat and the forest's relatively meager capacity to supply mammalian biomass. As was plainly evident, the collapse of wildlife populations was bringing an impoverishment of biodiversity and the decreased ability of local people to sustainably utilize wildlife for subsistence and as an important source of revenue.

In light of this, conservationists from the WCS Congo program focused attention on the production forests of the Kabo and Pokolo concessions to the south of the Nouabalé-Ndoki National Park. Instituting the necessary change in what at the time was an enormous, uncontrolled, illegal industry was challenging. Obtaining the legal mandate and practical wherewithal to work with the Congolaise Industrielle des Bois, local communities and government officials required both vision and inventiveness. Resistance to change was enormous, particularly among those who were heavily profiting from the illegal trade.

By dint of hard work, coupled with a grasp of the socio-economic foundations of the trade, the first functional system for the control of illegal bushmeat in a Central African forestry concession was launched.

The results of collaboration with stakeholders, enforcement of laws and the promotion of alternative protein sources were quickly apparent. Before the WCS conservation project was established in Bomassa and the subsequent creation of BZP, huge expanses of production forests could be traversed with only rare sightings of wildlife. After just a few years of recovery from the effects of unregulated hunting, forests and the natural clearings were again revealing the presence of elephants, gorillas and diverse ungulates. Community hunting zones around small villages on the periphery of the concessions once again began yielding sustenance. Some villages, with the greatest compliance with hunting regulations, showed hunting returns that approached long-term sustainability (Eves & Ruggiero, 2000). A model was born: one with many parents among conservationists and local people, but also one with detractors who preferred short-term profit. The realization among local communities and a logging operator, that the otherwise inevitable pattern of destruction could be avoided and that biodiversity conservation was possible where utilization was regulated and common interests respected, was however taking root.

The Buffer Zone Project was founded and implemented on the basis of four principles: (1) partnership with the private sector for conservation; (2) landscape conservation to link protected areas with the sound management of the multi-use forests surrounding them; (3) a multi-pronged approach to management based on responses to diverse threats; and (4) research-based, adaptive management (Elkan & Elkan, 2005; Elkan *et al.*, 2006; Poulsen, 2009).

Private-sector partnership for conservation is the partnering of companies and conservation organizations, public agencies and/or local communities to promote adoption of business practices that conserve habitat and biodiversity and secure the livelihoods of local people dependent on natural resources. *Landscape conservation* is the idea that protected areas are typically too small to ensure the long-term conservation of populations of wide-ranging animals and that considering social, cultural and economic needs is critical to conserving biodiversity in lands outside of protected areas (Box 1.6). Landscape conservation can enhance the conservation success of parks, reserves and the biodiversity they are meant to protect by addressing threats from human activity in the surrounding multi-use forests. A *multi-pronged approach* to conservation is one that recognizes that the threats to biodiversity are diverse and that no single action (silver bullet) can mitigate them all: law enforcement must be combined with awareness raising and the promotion of alternative economic activities. *Research-based, adaptive management* is the concept that we need to learn by evaluating how well or how poorly we are doing over time. Threats to biodiversity will evolve; adaptive management structures must be in place to allow conservation actions to evolve with them. Management initiatives must therefore be coupled with research and monitoring programs that facilitate the identification of new conservation threats and challenges and provide solutions to meet each challenge. Monitoring efforts also serve as an important means to verify the successes and failures of management activities.

Box 1.6 **The WCS landscapes species approach to conservation (David S. Wilkie, Wildlife Conservation Society)**

Getting the spatial scale right

If we want to conserve biodiversity and all other ecosystem services within an area, we need to act at a spatial scale that matters to the species that rely on that area. To do this, we first need to be explicit about what we are committing ourselves to conserve and understand clearly the habitat needs of these species and the threats to them.

The Wildlife Conservation Society has developed an approach to defining the spatial extent and configuration of landscapes and seascapes focused on wildlife, which is sufficient to ensure that functional ecological relationships among species remain intact over the long term. This landscape-species approach facilitates the conservation of full assemblages of native plants and animals within an area.

Our landscapes-species approach focuses conservation actions on a suite of marine or terrestrial wildlife species that together rely on all primary habitats and are adversely affected by all key threats to biodiversity in the area. These landscapes and seascapes species make conservation goals explicit. By protecting habitats and abating threats, we can create a 'conservation canopy' that confers protection to most plant and animal species in the landscape or seascape, in addition to the target species.

To implement the landscapes-species approach we have developed a series of technical tools and accompanying training manuals, available on our website (http://conservationsupport.org/ResourceLibrary/tabid/4103/Default.aspx).

The following chapters describe the inception, implementation and lessons learned while applying each of these principles of conservation. Chapter 2 discusses the formal institutional organization of the BZP. It examines the roles of the CIB, MEF and WCS, as well as their motives for entering into the collaborative relationship. Chapter 3 describes how communities are integrated into the co-management approach of the project and the development of land-use planning and access regulations for landscape conservation. Chapter 4 explains the nitty-gritty of implementing a multi-pronged approach to conservation, taking a look at project activities such as law enforcement, alternative activities, community conservation and awareness-raising and adaptive management. Chapter 5 and 6 examine results of the research and monitoring activities that are the backbone of adaptive management. Chapter 5 highlights the biological research that has been conducted over the past several years, assesses the results to date and offers ideas of how biological monitoring can be integrated into forest inventories. Chapter 6 summarizes the socio-economic research conducted in the concessions, defining methodologies for collecting rigorous data and evaluating the impact of logging on human communities. Chapter 7 presents the lessons learned to date from the project, in view of replicating it across other industries or in other forestry concessions in the tropics.

2

Building Partnerships for Conservation

John R. Poulsen and Connie J. Clark

Nicholas School of the Environment, Duke University,
Durban, NC

In an increasingly globalized world, the private sector is driving the global economy and, by extension, the environmental impacts of economic activity. The growth in extractive industries conveys potentially serious threats to the environment (Fearnside, 2001; Nepstad *et al.*, 2006; Butler and Laurance, 2008). The case of oil exploration in Nigeria is perhaps the most notorious example of environmental degradation associated with an extractive industry in Africa; many other examples exist. Alluvial mining of diamonds in places like Sierra Leone and Angola is associated with the devastation of gallery forests and aquatic fauna; in Zambia, copper and uranium mines threaten rivers with heavy metals and acidic discharges that have negative consequences for the health of local inhabitants and wildlife. In Senegal and Mauritania, international fishing fleets endanger artisanal fisheries.

Growing demand for Africa's natural resources means that money is available for exploration and extraction. It also means more money can be made available for responsible practices however, particularly as public concerns about environmental conservation and climate change put pressure on companies to protect natural ecosystems and the services and products they provide. One way to access and make use of these financial resources is through the development of private-sector partnerships for conservation (PSPCs). PSPCs are alliances between private companies and conservation organizations, public agencies and local communities to promote mutually beneficial and ecologically and socially responsible activities. Such alliances could take on many

Tropical Forest Conservation and Industry Partnership: An Experience from the Congo Basin, First Edition.
Edited by Connie J. Clark and John R. Poulsen.

different forms. The key is that business provides at least a portion of the financial resources to support conservation through partnerships with organizations that possess the expertise and tools to do conservation. PSPCs present an unprecedented opportunity to influence the practices of extractive industries and make them beneficial for both the company and biodiversity conservation.

PSPCs will be most effective if they target unnecessary and avoidable impacts, especially wildlife harvest in logging concessions. Some destructive practices in logging are, of course, unavoidable. In those cases, restoration may be the best option. Where restoration is not possible, the best recourse may be for the industry to pay for conservation offsets in the form of land set aside for conservation[1]. On the other hand, unnecessary and avoidable impacts such as bushmeat hunting must always be addressed.

Logging operations in the tropics increase pressure on wildlife by paying relatively high wages, thereby swelling the local economy and attracting large numbers of people (workers, family members and traders) into areas that were formerly sparsely populated (Wilkie & Carpenter, 1999). Once roads provide access to markets, bushmeat becomes a marketable commodity and hunting becomes transformed from a subsistence activity to a commercial one. Commercial bushmeat hunters often disregard land tenure and prior resource claims. Demand for bushmeat from logging company employees drives this market, because most companies fail to provide their workers with adequate animal protein. The gap in demand is met by local hunters, immigrants drawn to the boom economy and even company employees who supplement their logging income with bushmeat revenue using the company's roads and vehicles.

The Buffer Zone Project was created to mitigate the non-obligate impact of logging in timber concessions around the Nouabalé-Ndoki National Park. The three partnering organizations[2] – the Congolese Ministry of Forest Economy (MEF), the logging company Congolaise Industrielle des Bois (CIB) and the Wildlife Conservation Society (WCS) – brought together their financial, material and human resources to form an independent project to (1) protect the Nouabalé-Ndoki National Park from hunting pressure; (2) manage wildlife in the CIB logging concessions; and (3) mitigate the potentially negative effects

---

[1] Programs such as the Business and Biodiversity Offsets Program (BBOP) have developed principles and methodologies to support best practice in voluntary biodiversity offsets as a way to achieve no net loss of biodiversity relative to development (http://bbop.forest-trends.org/index.php).
[2] Safari Congo, a South African safari hunting company, was a fourth member of the partnership and signatory of the project protocol. The company ran bongo-hunting tours in the Kabo concession, but left the country (and the partnership) when the newly installed government clarified that bongo hunting was illegal.

of logging on the livelihoods of local people[3] (PROGEPP, 1999, 2008). With the goal of conserving wildlife and biodiversity, these partners broke away from traditional approaches to forest management that focused only on timber production. Working with local communities, the BZP implemented a multi-pronged approach to wildlife management that involved law enforcement, community-based conservation, development of alternative activities and adaptive management.

This chapter uses the BZP case study[4] to describe the necessary elements for building successful PSPCs that encourage environmentally appropriate resource use and wildlife and biodiversity conservation. It focuses largely on the institutional organization of the project, specifically the pros and cons of the partnership and the motivations and roles of the actors. In addition to describing how BZP works, this chapter assesses the characteristics of the partnership that fostered its success over the last decade. While some of the experiences from the BZP may be specific to logging and to Central Africa, the lessons learned about the components of an effective collaboration for biodiversity conservation and long-term resource use should extend more generally to a wide variety of industries and types of partnerships.

The following section examines the pros and cons of PSPCs, including the risks to which organizations expose themselves through partnerships. Examples are borrowed from the BZP to identify the characteristics of an effective partnership, emphasizing the necessary institutional components and the creation and maintenance of institutional relationships. The circumstances that motivated CIB, MEF and WCS to form the BZP are then described. The roles of CIB, MEF and WCS in the partnership, the activities each are involved in and how the participation of each organization evolved over time are discussed. The role of local communities in the management and conservation activities is also examined. The penultimate section (before conclusions) assesses some of the formal and informal attributes that have allowed the BZP to move forward, often in the face of institutional conflict.

---

[3] Although local communities are not signatories of the protocol creating the collaboration between MEF, CIB and WCS, the BZP works with and alongside local communities as well as other governmental and non-governmental agencies (Chapter 3).

[4] The possibilities for partnership with the private sector are endless: working with ecotourism companies to monitor wildlife and educate tourists, collaborating with local transport companies to find ways to reduce or stop the transport of bushmeat and partnering with logging or mining companies to manage wildlife over hundreds or thousands of hectares of forest.

Pros and cons of private-sector partnerships for conservation

There is a growing recognition that working with industry can be an effective way to conserve biodiversity (Cyranoski, 2007). To succeed, differences between the partners in goals and approaches must be reconciled. The first step is to understand the goals of each partner and their motivations for entering into a partnership. A high-level manager in CIB once gave the following advice: "Private enterprise would take on board any environmental measure as long as it was convinced that it makes business sense. Therefore try to understand what motivates private enterprise and be sympathetic to their concerns so that you can demonstrate the environmental as well as the financial benefits to them". Some of the pros and cons of partnering for the private sector and NGOs or other public agencies are discussed in the following (see Box 2.1 for pros and cons for government), illustrated with examples drawn from the BZP.

Box 2.1 **Government membership in a PSPC**

Like business and conservation organizations, government can also be a member of a PSPC (e.g., BZP). Its role will largely depend on the national context of the country, including the technical capacity of its personnel, the wealth of the country and the level of democratization. Membership in a PSPC has benefits and risks for government (Elkan & Elkan, 2005; Poulsen, 2009).

Pros for government:

1. financial and logistical assistance in managing wildlife and enforcing laws;
2. technical assistance in monitoring wildlife, impacts of industry, etc.;
3. greening its reputation;
4. tax revenue, infrastructure development and employment gained from environmentally and socially responsible resource exploitation; and
5. environmental and social benefits of resource conservation.

Cons for government:

1. pressure to improve standards or investment in conservation;
2. criticism for not doing its share (case of developing countries without the personnel, technical capacity, resources and/or political will to act as an equal partner in the PSPC);
3. loss of national autonomy to 'neo-colonial' pro-West NGOs;
4. sacrifice of national good and natural resources to business interests.

The pros of partnering with NGOs

The private sector usually contributes to biodiversity conservation because it is good for the bottom line. In conservation organizations, companies gain a partner that can help them meet national and international environmental standards, obtain access to financial resources and meet needs that the company is unable to meet by itself (or can only meet with great investment and cost).

Complying with legal standards

Laws may require industry to take measures to reduce or mitigate its environmental impact on biodiversity. In Congo, for example, national laws require logging companies to contribute to the protection of wildlife by paying for eco-guards to enforce wildlife laws. The goals of a logging company are to produce timber and make profits, and managing eco-guards and wildlife is outside of its expertise. Partnership with WCS made compliance with government standards easier for CIB.

Projecting a 'green' image

By partnering with a well-known NGO, a company can improve its reputation and justify its presence and its methods. As CIB expanded its operations in northern Congo, critics accused the company and its employees of killing apes (World Rainforest Movement, 2003). Partnering with WCS blunted those attacks by demonstrating that the company was willing to collaborate for conservation, thus 'greening' its public image.

Expanding access to markets

Increasingly, some markets (notably in European countries) require that products such as timber and fish be managed and harvested according to environmental standards that ensure quality and protect biodiversity, productivity and ecological processes. Partnering with an environmental organization can help the company to meet such standards obtain certification and access premium markets where their products may be sold at higher prices.

Opening access to financial resources

Access to multilateral funding agencies such as the World Bank is conditional on meeting environmental and social safeguards. Partnering with a conservation NGO demonstrates a commitment to good environmental practices, particularly because the NGO is expected to 'blow the whistle' if the company does not meet its commitments. In addition, companies can also gain access (at least indirectly) to their partner's donors. Through its donors, including US Fish and Wildlife Service (USFWS), Liz Claiborne Art Ortenberg Foundation, International Tropical Timber Organization and others, WCS raised millions of dollars for conservation of wildlife in the CIB concessions. All those funds indirectly helped the company by allowing it to meet environmental standards to promote the sale of its timber products.

Maintaining consistency in financial resources for conservation

Companies and NGOs both have fluctuating economies, but they do not tend to fluctuate perfectly in phase. Partnering can reduce the number of disruptions to biodiversity conservation and management as one partner provides the funds to maintain activities when the other is undergoing an economically lean period.

Conserving resources

Companies that derive profits from renewable resources such as fisheries, timber, safari hunting and ecotourism have an interest in conserving the ecosystems that produce those resources. Companies may also benefit from the conservation of resources from which they do not directly profit. For example, while CIB derived revenue from the production of timber, the dependence of its employees on wildlife and freshwater fish as principal sources of protein spared the company considerable costs (Poulsen *et al.*, 2009). Without natural resources, the company would have had to invest greater time and money into providing meat for its personnel (Poulsen *et al.*, 2007).

Benefiting from conservation expertise

Most conservation NGOs have knowledge gained from years of experience. Subjects of importance to conservation biology (population biology and genetics, community ecology, animal behavior, wildlife management, environmental education, experimental design and statistical analysis) may not always find their way into the log yard or the boardroom. Companies can tap into knowledge and experience through partnerships with NGOs.

Building relationships with local communities

Relationships with local communities can be strained or even contentious when companies bring in large numbers of outsiders, create disruptions such as truck traffic, and exploit what locals believe to be theirs. Many companies compensate communities through community relations and development (e.g., giving uniforms to a soccer team, building schools, providing jobs or environmental clean-up). PSPCs can avoid or mitigate conflict by working with local people to manage their resources. Conservation projects such as BZP also build capacity and provide jobs (e.g., forest guides, eco-guards, etc.) for less educated members of the community that may not be well qualified for jobs with the industry.

The pros of partnering with business

Conservation organizations will partner with the private sector to the extent that such partnerships contribute to biodiversity conservation. Adding a private-sector partner to the equation can provide much-needed resources and legitimacy.

Accessing financial and logistical resources

Private-sector partnerships are usually based on an exchange of financial and logistical support for resource management and conservation. For NGOs and public agencies with tight budgets, these resources can be critical to their effectiveness and sometimes their existence. Private-sector money often comes with fewer restrictions than government or donor funds. For example, whereas US government funds (e.g., USFWS) cannot be used to pay eco-guard salaries, private logging companies rarely have such restrictions. By the mid-2000s in northern Congo, several logging companies started to pay eco-guard salaries to enforce wildlife laws. Logistical support can also be an important contribution in remote areas where access is difficult: CIB provided electricity, fuel and electrical and mechanical services to BZP, which were logistically challenging to procure without the company. Logistical support may be provided on a pay-for-services basis, but without the company the services would be unavailable.

Gaining access to areas of high biodiversity value

Industries usually lease the concessions they exploit, particularly in the mining, logging and petroleum sectors. Leases bestow rights and authority *vis à vis* the government and local communities, sometimes conferring companies with the power to decide which stakeholders work in an area and what they are allowed to do. Conservation organizations can therefore gain access to areas of high biodiversity value that could otherwise be off limits or in which their roles could be largely restricted (Elkan & Elkan, 2005).

Planning for the future

In some cases, industry may stay in an area for long periods of time. Selective logging in northern Congo, for example, was planned on a 30-year rotation (Congolaise Industrielle des Bois, 2006). A company that plans its operations over a long time horizon is more likely to make investments in the long-term management and conservation of an area than a company that plans to exploit and exit. Most donors to conservation projects give money on a 1–3 year time-frame, even though managing threats to biodiversity usually takes decades. In a world where environmental donors tend to fund the newest, trendiest issues, the financial and logistical support from the private sector may be the best opportunity for sustaining biodiversity conservation over a span of many years.

Providing authority

A formal partnership with industry grants the conservation organization or project with authority in the eyes of company employees. Even with governmental permission to work in the same area as industry, conservation projects can be treated as a nuisance or perceived as competing with a company that provides greater opportunities for jobs and salaries. Acceptance by the company leadership, and the image that the company and conservation organization work together, legitimizes conservation management and makes it easier to accomplish.

Improving resource management from the inside

A formal partnership with the private sector confers access to company leadership. Both formal and informal interactions with company personnel can dramatically influence the company's policies and actions towards the environment and biodiversity. Informal conversations between BZP personnel and CIB crew leaders about growing incidences of elephant poaching led crew leaders to crack down on their own employees.

Improving conservation at the conference table

Through PSPCs, conservation organizations acquire a seat at formal meetings with industry, government and other stakeholders. Such meetings offer an opportunity to influence policies and standards. During the development of management plans for the CIB concessions, WCS participated in meetings with CIB and the government to decide logging techniques and road construction standards. In these meetings, WCS expressed a conservation perspective and infused science into policy discussions so that decisions were based on solid, technical information. Data-supported science-based arguments for management and policy decisions hold much greater weight than beliefs or positions.

Potential risks of partnering with industry or conservation organizations

The benefits of partnering with industry or conservation organizations often come with a cost. Many of the potential risks are shared by both partners and include the following.

Taking on financial and logistical burdens

As discussed previously, the private sector brings logistical and financial resources to a PSPC. Depending on the level of commitment, these costs can be considerable. Moreover, the contribution can grow over time either because the industry's impact on the environment is more damaging than originally expected, national standards grow more rigorous or the breadth of activities undertaken by the partnership increases.

Sleeping with the enemy

Conservation organizations that partner with the private sector may be perceived as 'sleeping with the enemy'. They may be accused of lowering standards or compromising ethics for financial gains. If this is the case, it is the burden of the conservation organization and PSPC to prove critics wrong.

Receiving bad press

Businesses, NGOs and government agencies sometimes make bad decisions, suffer poor management or fall into problems of corruption or poor performance. If any of these problems inflict a PSPC, the 'innocent' partner can be implicated and may receive bad press. For an NGO, bad press will hurt its reputation, ability to work elsewhere and ability to raise funds. For businesses, bad press will not only hurt its public image but could also affect sales, profits, perception of risk and the ability to raise capital.

Arguing among 'friends'

Companies and conservation organizations do not usually have the same goals and will not always see eye-to-eye. While effective PSPCs have mechanisms for resolving conflict, partnerships can turn bad for unforeseen reasons. At a minimum, resolving conflict entails an investment of personnel, time and energy from all involved organizations. Failure to invest in the partnership may result in a broken partnership, bad relations, bad press and organizational conflict.

Inviting pressure to improve standards

By partnering for biodiversity conservation, companies are opening their doors to conservation organizations and inviting them to the conference table. As such, the NGO can put pressure on a company to go further in its environmental standards than it had originally intended, at increased expense.

Inviting pressure to compromise standards

Just as industry invites pressure to improve its standards at increased cost to the company, conservation organizations invite pressure to compromise their standards. Sometimes this may be healthy; unrealistic standards can be softened through improved understanding of competing interests (social, cultural and economic) and through negotiation. However, NGOs or public agencies that turn a blind eye to unethical or illegal practices or lower their standards too far will compromise their reputations and reduce their effectiveness.

Guilty by association

Being viewed as a legitimate stakeholder by company employees and local communities is one of the benefits of participation in PSPCs for conservation organizations. However, if local communities come to view either the company or conservation organization in a poor light, their disapproval may be shifted to the partner.

Banking on continuity

Any number of disruptions in the structure of one of the member partners could compromise the resources, time and effort invested in a PSPC. Turnover in staff could alter interpersonal and organizational relationships on which the PSPC was founded. Larger disruptions such as turnover in government or change in company ownership[5] could also have major impacts on the partnership.

Motivations of MEF, CIB and WCS to form the BZP

On any given conservation landscape, a large number of stakeholders may have a vested interest in, or could be affected by, conservation decisions. This complexity notwithstanding, not all stakeholders will be critical actors.

---

[5] The global recession of 2007–2010 slowed demand for timber and challenged the ability of CIB to meet its financial goals. As a result, DLH sold CIB to Olam International Limited in 2010. The future of the BZP PSPC will depend on Olam's continued commitment to FSC certification and wildlife conservation.

The type of management system and the roles that need to be filled to affect conservation will determine the actors. These should be the actors that have the capacity, power, mandate and motivation to undertake these roles (Castillo *et al.*, 2006). The incorporation of actors into a partnership for conservation should ideally be formally assessed, but sometimes there may be little choice when it comes to selecting partners. For conservation organizations, the decision to partner with industry will be informed primarily by the conservation value of an area; characteristics of the potential private-sector partner will be a secondary consideration (see de Queiroz *et al.*, 2008 for specific criteria to consider when vetting a company for a partnership). For industry, the economic value of the site or activity will more seriously influence these decisions. In the case of the BZP, the partnership was one of convenience rather than an intentional strategy: CIB and WCS both worked in the area, and conservation demands pushed them together (Box 2.2).

Box 2.2 **Creation of the BZP (Jerome Mokoko-Ikonga, Assistant Director, WCS-Congo)**

The creation of the BZP was the result of a failure to create a zoning system around the Nouabalé-Ndoki National Park. In April 1998, WCS submitted a zoning plan to the MEF and to CIB that proposed setting the concession areas bordering the park off limits to logging and hunting. The government rejected the plan. Concerned about the impact that industrial logging could have on the park, however, the government, CIB, Congo Safari and WCS agreed to a system of management and law enforcement in the timber concessions adjacent to it. The management system was meant to stop poaching and bushmeat trade that would come from the installation of new worker camps, the opening of roads and the immigration of job-seekers. In June 1999, a historic protocol was signed among the four partners. This was the first such agreement in Africa establishing a collaboration between those who exploit the forest and those who conserve it. CIB's FSC certification was largely the result of its involvement in conservation. This project, which has overcome many obstacles to reach its thirteenth year in 2012, is a cause for optimism for conservation.

The initial motivation of WCS to participate in the BZP came from its goal of protecting the Nouabalé-Ndoki National Park (Box 2.3). The park was arguably one of the most remote and least-disturbed protected areas in the world and was one of WCS's flagship projects. CIB's motivation came partly from the philosophy of its founder, Dr Hinrich Lueder Stoll, as well as the desire to improve its reputation and to seek certification for its timber operations (Box 2.4). While MEF likely shared some of the same interests and motivations as WCS and CIB, it was also motivated by a desire to improve the livelihoods of the people in northern Congo through employment opportunities and the development of new livelihood activities. At least three characteristics of this partnership emerged to enable the successful establishment of the project: a foundation of trust between WCS and CIB managers, shared objectives and the complementary capacities and expertise of the partnering organizations.

Box 2.3 **WCS's perspective on the BZP partnership (Paul W. Elkan, BZP Project Director 1999–2003; Director WCS-Congo 2003–2007; Sarah Elkan, BZP Manager 1998–2003; Financial Director WCS-Congo 2003–2007)**

The creation of the BZP was the culmination of efforts initiated in the early 1990s. Recognizing the ecological importance of the area, the Government of Congo (with technical support from WCS and funding from USAID) changed its status from a timber concession to the Nouabalé-Ndoki National Park in 1993. Because conservation of the park (and of wildlife that ranged outside park boundaries) would be directly impacted by the surrounding logging concessions, the decree gazetting the park also called for the creation and sustainable management of buffer zones surrounding it.

In the early 1990s, the Mokabi and Loundougou concessions to the north and east of the park were not yet leased to logging companies. Kabo to the south had been logged in the 1970–80s by the *Société Nouvelle des Bois de la Sangha* (SNBS), but the company was going bankrupt and by the early 1990s its operations had been greatly reduced. Logging operations in Kabo had already led to the establishment of large commercial bushmeat and ivory poaching networks, however. Further to the south Pokola had been logged by the Congolaise Industrielle des Bois (CIB), beginning in the late 1960s. It too was subject to extensive bushmeat hunting and trade to feed logging crews and to support its growing logging town (also called Pokola). Hunters and traders used CIB's camps and vehicles to hunt bushmeat and to poach elephants for ivory, often with the direct involvement of CIB employees. Pokola was a harbinger of the impacts that would threaten other concessions and the park if additional protective measures were not taken.

From the beginning, the Nouabalé-Ndoki project sought not only to establish the park, but also to work with the government and private timber companies to promote wildlife conservation and responsible forest management in adjacent concessions. To that end, WCS sought to develop a constructive dialogue with companies in the area. Under the leadership of Dr Hinrich Stoll, CIB was open to discussion; Dr Michael Fay of WCS built a relationship with the company to promote improved forest management.

By 1995, international groups were criticizing CIB for its role in hunting and ivory poaching. In response, CIB invited several conservation and development organizations to visit its concession. This resulted in a series of recommendations to address the bushmeat trade and improve forest management. Unfortunately, few concrete actions were taken and the situation remained largely unchanged when civil war broke out in June 1997, causing most organizations to flee the country. In addition to recommendations, CIB needed practical technical assistance but few organizations were willing to work in the field with the company to address conservation problems.

CIB acquired rights to Kabo and Loundougou in 1997, rapidly expanding its infrastructure and road network despite the ongoing war. The company built a dike across the previously isolated Ndoki River and established a logging camp just 30 km south of the park, opening access to frontier forest. With the creation of infrastructure and roads in the Kabo concession, the expansion of operations compounded old pressures on wildlife (e.g., revitalizing elephant poaching networks) and greatly increased new pressures (e.g., systematic export of wildlife).

WCS sought to work with CIB as it expanded into Kabo and Loundougou, and proposed several alternative scenarios for the placement of infrastructure and roads to reduce their negative impact and to establish a buffer zone around the park. In early 1998, WCS proposed a buffer zone plan around the park that would create buffer zones in the Kabo, Loundoungou and Mokabi concessions and extend the park borders to include the pristine Goualougo area.

The Ministry of Forest Economy (MEF) rejected WCS's plan, prioritizing economic concerns over conservation. However, the government also recognized the importance of WCS's proposal to integrate wildlife conservation into forest concession management and requested that WCS develop such a program with the government and CIB.

Over the course of a year, the parties negotiated an agreement to establish a multi-faceted program, which was made possible by the common goals and trusting relationships developed among individuals from CIB, WCS and the ministry. Frederick Glannaz, CIB Management Plan Coordinator and son of the CIB Operational Director Jack Glannaz, was highly supportive of collaboration with WCS. Dr Marcellin Agnagna, Director of Wildlife for the MEF, worked closely with WCS and CIB to establish the BZP. In June 1999, CIB, MEF and WCS signed an agreement establishing the *Projet de la Gestion des Écosystèmes Périphériques au Parc* (PROGEPP).

Signing an agreement was one thing; implementing a field conservation program in a logging concession with a long history of rampant commercial poaching was another. Practical approaches on the ground and relationships of mutual trust and respect among key staff of the three organizations made the project successful. In addition, whereas most organizations pulled out of Congo because of the civil war, WCS and CIB remained. This also contributed to the partners' synergy in the early years of the program.

Addressing the widespread, lucrative and entrenched commercial bushmeat trade was a delicate and, at times, dangerous joint undertaking. WCS's Antoine Moukassa and CIB's Jacques Glannaz were instrumental in gaining the support of local communities and CIB employees. At times, the partners came under intense local pressure (e.g., CIB employees threatened to strike in protest of restrictions to bushmeat trade and poachers made death threats to WCS staff) and international pressure (e.g., an international environmental NGO publicly criticized the program). Facing these pressures together strengthened the partnership.

WCS stayed the course because it was able to gauge its progress in conserving wildlife populations in the concessions and within the park. WCS was also encouraged by the degree of involvement and support from the most senior levels of CIB and the government. For example, Dr Stoll personally participated in meetings with the workers' union to negotiate the adoption of wildlife management regulations as part of the CIB's internal rules. Dr Francois Ntsiba, Director General and subsequently Director of the Center for Inventories and Forest Management of the MEF, supported the integration of BZP's methods into forest concession management throughout the country.

The BZP achieved success in hunting management, wildlife protection, conservation awareness and establishment of set-aside protection zones. It fell short of WCS's objectives in terms of road placement and sawmill construction issues. The program was instrumental in forest concession management planning and FSC certification processes. The park received buffer protection and several key elements of the original zoning plan have been incorporated in concession management plans. The program was the first field-based initiative in Central Africa to effectively incorporate wildlife conservation principles in forest concession management and provides valuable lessons for similar initiatives elsewhere in the Republic of Congo and the Congo Basin.

Box 2.4 **CIB's perspective on the PROGEPP partnership (Lucas Van der Walt, CIB Environmental Coordinator 2001–2010)**

Congolaise Industrielle des Bois (CIB) is recognized as a model for sustainable forest management. Despite its much-publicized achievements over the last 10 years, including the BZP partnership, CIB's journey started years earlier. CIB's dedication to the BZP was born not only out of its founder's vision but also from individuals willing to work together in the field and to find compromises for the greater good.

CIB's German founder, Dr Hinrich Lueder Stoll, studied forestry under Professor Franz Heske at the University of Hamburg shortly after World War II. He studied the principles of Sir Dietrich Brandis, recognized as the father of tropical forestry. Brandises' teachings had a profound impact on Stoll, as they did for the founder of the US Forest Service Gifford Pinchot. By the 1960s, Stoll had founded timber procurement offices in Central and West Africa; however, he had yet to establish a forestry company where he could implement Brandises' principles. In 1967, he bought two companies in the Congo (Brazzaville) and merged them into a single company: CIB. Until 1997, CIB managed the Pokola concession.

By the late 1980s, Stoll had started social and forest management initiatives using the company's financial means. He met WCS's Mike Fay, but Fay was focused on the Kabo concession whereas CIB only managed the Pokola concession to the south of Kabo. At the time the idea of working with a logging company was also too politically sensitive for WCS. Stoll sponsored the doctoral research of Hermann Fickinger on the sustainability of harvesting timber species in Pokola. A second doctoral study by Jerome Lewis, although not directly sponsored by Stoll, served to highlight the plight of the local Mbendzélé[6]. These studies motivated Stoll in later years to seek out expertise on issues outside of core company operations.

Due to pressure from international environmental NGOs, a meeting in Pokola with local stakeholders and the government led to the first CIB policy on wildlife management in December 1995. The declaration served as the basis for the eventual 1999 agreement between the BZP partners. Sustainable forest management was a key theme in the 1995 document and it was an important step to making the government and stakeholders in Pokola aware of the concept. The Congolese government, WCS and CIB all played vital roles in the successful implementation of BZP; without the other partners it is doubtful that CIB would have continued to integrate wildlife and forestry management on its own.

---

[6]Mbendzélé are forest-dwelling semi-nomadic ethnic minorities indigenous to northern Congo. They are a clan of the Aka peoples, sometimes referred to as Pygmy.

The civil war (1997–1999) in Congo, which saw most international organizations and companies pulling out of the country, established a regular contact between CIB and WCS staff that stayed the course. Stoll played an important role in laying the foundations for CIB's involvement in BZP; however, it was the unique friendship between Fred Glannaz (son of CIB's operational director) and Paul Elkan (WCS researcher) that largely contributed to its success. Good relations between individuals often proved vital for the continuation of the project. As much as CIB's motivation to enter the agreement came from Stoll, its ultimate success came largely from the CIB and WCS team on the ground. BZP was CIB's first multi-stakeholder project and served as a catalyst for its future progress in sustainable forest management. Today CIB continues in the path laid out by its founder; it obtained Forest Stewardship Council (FSC) certification in 2006. Its example spurred many others in the industry to follow; today tropical Africa boasts the largest area under FSC certification in the tropics.

Foundation of trust

The early success of BZP was largely attributable to the personal relationship and transparent communication between the WCS project director, Paul Elkan and the CIB forestry manager, Fred Glannaz (Elkan & Elkan, 2005). Their friendship and shared ideas contributed to the development of a wildlife management system, while largely avoiding the organizational conflict that may have been expected between a logging company and a conservation organization. Years later, when differences in approach between WCS and CIB strained the partnership, this relationship was sometimes invoked as a reminder that the two organizations could work closely together. Strong personal relationships have proven to be significant in building strong communication and breaking the barrier between 'them' and 'us'.

Complementary institutional capacity and expertise

For each management structure, there is a set of management roles and activities that are likely to affect conservation of wildlife and natural resources in positive ways (Castillo *et al.*, 2006). The most effective mix of actors is one in which the quality of the actors' attributes matches the specific needs of the management system. The BZP partners have proven to be compatible in the sense that they created an effective blend of *institutional mandate to manage, motivation for conservation, power to influence* and *capacity to act*. Together they have the legal mandate and authority as well as the financial resources and technical know-how to influence conservation problems.

Same objective, different goals

Despite having vastly divergent institutional goals and different short-term motivations for venturing into the BZP, the partners shared the same long-term objective. Wildlife management is both a way to preserve biodiversity and endangered species and is a way to conserve a source of wild meat for local people. This objective clearly fit with WCS's institutional goal of conserving wild places and wild animals. It did not clearly fit CIB's goal to produce timber and wood products to make a profit for its shareholders. Nevertheless, wildlife management complemented profit making to the extent that it permitted the company to build its reputation as a responsible, environmentally friendly company. The goal of the government was to develop its economy and infrastructure; wildlife management fits this goal by providing food security for local people, generating employment opportunities through wildlife management and ecotourism and by improving the government's image with local communities and international donors.

 The greatest conflicts among partners came when there were perceived differences in the objectives of the PSPC. The MEF's greatest criticism of the project over the years was the very modest result of the alternative activities program. The alternative activities program was meant to identify and develop livelihood options, such as animal husbandry, forest-appropriate farming and agriculture and even small business development. While the government envisioned the activity as a type of rural development program that would generate real income for families, WCS viewed it as a way to produce alternative sources of protein in order to reduce hunting and the bushmeat trade. As such, WCS encouraged CIB to make domestic sources of protein available to its employees and their families, rather than allowing them to subsist on bushmeat. WCS also held the view that CIB should make a bigger commitment to training and employing indigenous people rather than immigrants because employment with the logging company was a readily available and lucrative source of family income. The BZP never had the budget or the expertise to implement the large-scale development program necessary to provide domestic protein and income for 20,000 people.

A more fundamental conflict occurred over the definition of the BZP goals. Whereas WCS defined the primary goal of the BZP as the protection of the Nouabalé-Ndoki National Park, CIB defined its primary goal as wildlife management within its concessions. At the center of this issue was the question of where the majority of resources would be allocated. WCS viewed the biggest gains in biodiversity conservation as coming from protecting the park from incursion by poachers. For its management plans and Forest Stewardship Council (FSC) certification, CIB needed a wildlife management and protection strategy that covered the entirety of its concessions. As long as the Nouabalé-Ndoki National Park remained largely free of poachers, the BZP continued to operate over the entirety of the concessions; an upswing in incidents of poaching in the park could however have resulted in an organizational conflict if law enforcement resources were shifted to concession areas close to the park borders.

Roles of the BZP partners

Having discussed the incentives to partner for conservation, we now examine the institutional roles of CIB, MEF and WCS in implementing the project. As mentioned previously, conservation is most likely to be affected by actors with the necessary mandate, motivation, power and capacity. This was the case within the BZP. As a result, the degree to which an organization participated in any aspect of the project activities varied with the activity and the ability of the organization to contribute to its implementation and success (Table 2.1).

MEF was officially responsible for enforcement of hunting and wildlife laws, which involved the management of the eco-guard unit. The MEF also represented the project in case of litigation and served as the liaison to other governmental departments. In addition to law enforcement, the MEF contributed to the planning, implementation and oversight of all other project activities.

WCS was responsible for implementation of most project activities with the exception of law enforcement, for which it provided technical support. WCS took the lead role in environmental education, community conservation, research and the monitoring of conservation activities, wildlife populations

Table 2.1 **Partitioning of activities based on capacity and mandate of the BZP partners.**

Category	Actor	Activity	Capacity/Mandate
Community outreach	MEF	Village meetings to open the legally recognized hunting season (May–October)	Knowledge of hunting laws; Mandate to enforce hunting and wildlife laws
	WCS	Formal classes in schools; Awareness-raising campaigns in villages and logging towns; Distribution of posters and pamphlets in villages and logging towns	Expertise in wildlife management and conservation; Experience in education and awareness-raising; Access to international donors
	CIB	Distribution of posters and pamphlets to employees; Transmission of conservation movies and radio shows (usually produced by WCS collaborators) on company TV and radio stations	Logistical means to build and operate TV and radio stations; Mandate to inform employees of company regulations and ethics
Law enforcement	MEF	Management of eco-guard unit; Training of eco-guards; Procurement of weapons	Mandate to enforce hunting and wildlife laws
	WCS	Technical and strategic advice on law enforcement and eco-guard management; Funding of law enforcement effort and performance; Funding of eco-guard missions	Expertise in wildlife management; Expertise in project management; Access to international donors

(continued overleaf)

Table 2.1   (*Continued*)

Category	Actor	Activity	Capacity/Mandate
	CIB	Funding and logistical support of eco-guard unit (salaries); Construction and maintenance of roadside posts; Enforcement of company rules (sanctioning violators)	Obligation to support wildlife law enforcement
Alternative activities	MEF	Appointment of Agricultural Technician to the BZP	Expertise in agriculture
	WCS	Experimental animal husbandry; Micro-projects to increase protein production	Expertise in community conservation; Access to international donors
	CIB	Importation of livestock and frozen meat into concessions; Maintenance of refrigeration chambers to conserve domestic meat and fish; Experimentation of fish culture through construction of fish ponds	Obligation to make protein available to its employees; Expertise in livestock and fish production
Research and monitoring	MEF	Grant of permission to project partners to conduct research in protected areas	Mandate to oversee management protected areas and forestry concessions

(*continued overleaf*)

Table 2.1   (*Continued*)

	WCS	Monitoring of large mammal populations;	Expertise in wildlife management;
		Monitoring of human demographics and livelihoods; Monitoring of bushmeat availability; Monitoring of law enforcement effort and performance	Expertise in research and science; Access to international donors
	CIB	Financial and logistical support of one-off wildlife surveys; Inventories of commercial tree species	Expertise in forest management
Reduced-impact logging	MEF	Review of RIL procedures and verification of good implementation	Mandate to manage forests
	WCS	Provision of technical advice on minimizing impact of logging on wildlife populations and review of all RIL procedures; Independently monitors logging practices and reports any breaches of procedure to CIB management.	Expertise in wildlife management
	CIB	Design and implementation of reduced-impact logging program	Expertise in forest management

and socio-economic activities. In practice, WCS historically played a large role in law enforcement by offering logistical support and technical assistance for the planning of law enforcement missions, management of eco-guards and monitoring of law enforcement results (Elkan & Elkan, 2005; Elkan *et al.*, 2006).

CIB was responsible for forestry-related activities such as implementing a reduced-impact logging program and developing and implementing management plans for its concessions. Most notably for biodiversity conservation, CIB enforced its company rules relative to wildlife management. It educated its employees about wildlife laws, penalizing them for infractions. The company also supported biodiversity conservation by contributing to the logistical and financial support of the eco-guard unit (paying approximately three-quarters of the annual cost), supplying housing for most of the BZP employees and providing relatively inexpensive logistical support (e.g., mechanics, electricians, etc.). To meet its obligation to reduce bushmeat hunting in its concessions, the company invested materials and manpower to increase availability of domestic protein for its workers and their families. Compared to other logging companies operating in Central Africa, CIB took unprecedented steps to integrate wildlife management into logging procedures, policies and land-use practices.

One way to understand the breakdown of roles and activities of these three organizations is to 'locate' them along axes that represent the qualifications needed to fulfill the priority roles in management. Borrowing from the framework developed by Castillo *et al.* (2006), each candidate actor (CIB, MEF, WCS and indigenous and immigrant communities) is represented as a point along four axes indicating their mandate to manage, capacity to act, motivation to conserve and power to influence relative to the other candidate actors at the site. We distinguish between 'immigrants' and 'indigenous' people to emphasize the role of industry in changing human demographics. 'Immigrants' are defined as people who migrated to the concessions from a different region of Congo or from a different country, whereas 'indigenes' are people from northern Congo, including both Bantu groups and Mbendzélé. It is important to keep in mind that identifying the position of any actor along these axes is a subjective judgment. These judgments are formalized and graphically displayed in radar diagrams that rank actors' perceived competence along a scale from 1 (low) to 5 (high) with high competence near the center (Figure 2.1). Drawing lines to connect the locations of an actor or organization along each axis results in a polygon that depicts the overall strengths of an actor to fulfill a desired role. Overlaying these diagrams for multiple actors for any given role enables us to review the relative strengths of different actors.

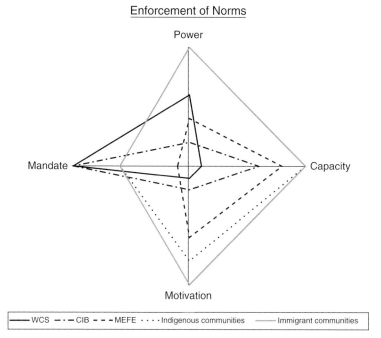

Figure 2.1   Enforcement of norms/wildlife laws was a priority management role for BZP. As depicted in this radar diagram, the appropriate mix of actors to enforce laws was MEF, WCS and CIB. Reprinted with permission from the Wildlife Conservation Society's Working Paper No. 28, 2006, pp. 70.

This process was conducted for three different roles: (1) enforcement of wildlife laws; (2) communication of wildlife laws; and (3) building constituency.

The enforcement of wildlife laws is an example of how organizational capacity and mandate stratifies the degree of involvement in different activities. Prior to the creation of the BZP, logging truck drivers facilitated the bushmeat and ivory trade by transporting wildlife products to markets (Elkan & Elkan, 2005; Elkan *et al.*, 2006). MEF had the sole official mandate to set and enforce access regulations and hunting laws, but lacked the capacity and motivation to do so effectively (although this improved with experience gained at BZP). On the other hand, WCS had the capacity and motivation to implement project activities; it brought both technical knowledge of wildlife management and access to funds from international donors to the table. Law enforcement efforts would however have failed if CIB had not allowed its vehicles to be searched and had not penalized or fired employees implicated in the bushmeat trade. Through the threat of disciplinary action, CIB wielded the necessary power to alter the behavior of its truck drivers. When united, CIB's power and motivation to influence employee behavior coupled with MEF's legal mandate and WCS's motivation and technical capacity enabled the implementation of law enforcement in the forest concessions (Castillo *et al.*, 2006).

Enforcement of hunting laws, however, conflicted with the cultural traditions of the indigenous populations living in the concessions. Hunting elephants, for example, was traditionally a 'right of passage', a source of income and favored protein source. Before the initiation of BZP, rural communities were largely ignorant of wildlife laws. Raising awareness of the laws served as an important step in the process of building support for conservation and compliance with wildlife laws. As depicted in the accompanying radardiagram, WCS possessed the strongest combination of attributes to facilitate this communication process (Figure 2.2).

In many situations where human population densities are low, hunting is restricted mostly to subsistence use and wildlife populations are productive and renewable. Under these conditions people can typically hunt wild game without drawing down wildlife populations to the point of collapse, even in open-access systems. Population growth (particularly the immigration of people for logging activities) and the commercialization of hunting can however exert heavy pressure on wildlife, leading to the decline and even the extirpation of wildlife populations. To prevent the depletion of legally hunted wildlife resources, a controlled access management regime was required to limit who can hunt and establish the quota per hunter. The establishment of such a system for legally hunted species required building support for conservation and the cooperation of local populations, most of whom had little understanding of the long-term consequences of their altered patterns of resource use.

The ability of actors to build a constituency in support of the conservation of populations of legally hunted species was strongly dependent upon their motivation and power to influence groups of local users. At conception of the project, the local users (both indigenous and immigrant) were experiencing short-term financial rewards from over-hunting, but not experiencing species declines serious enough to alter their perception that wildlife was an unlimited resource. They therefore lacked the motivation to serve as constituency builders for the conservation of wildlife. Because CIB employed much of the immigrant population, it possessed the power to influence immigrant communities through work-related incentives. WCS, on the other hand,

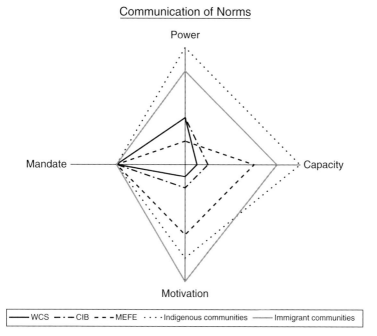

Figure 2.2 **Communication of norms and wildlife laws to rural populations was a second priority management role. In populated areas, compliance with hunting regulations required that people understand the laws protecting endangered species and how those laws are enforced. Capacity and motivation to plan, implement and finance awareness-raising activities were the priority attributes of appropriate actors to fill this role. Among potential actors to fulfill this role, WCS had the strongest technical capacity. © 2010 Wildlife Conservation Society. Reprinted with permission from the Wildlife Conservation Society's Working Paper No. 28, 2006, pp. 71.**

influenced indigenous populations because a high proportion of its staff originated in the Sangha province and could identify and communicate with indigenous groups. In addition, WCS staff had previously conducted social surveys and awareness-raising campaigns. Both CIB and WCS exhibited reasonably strong technical and logistical capacity and strong motivation to build support for biodiversity conservation, although the source of motivation differed greatly. WCS and CIB therefore worked together to build a constituency for the conservation of legally harvested species in the logging concessions (Figure 2.3).

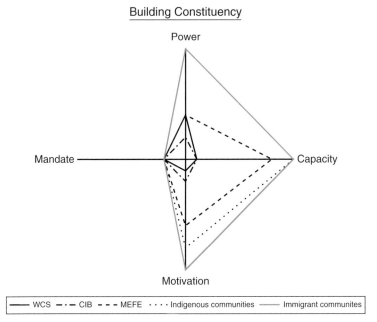

Figure 2.3   **Building a local constituency of support was a priority management role to conserve populations of legally hunted species in logging concessions.** Among potential actors, WCS had been the most motivated and dedicated to the cause of conservation. CIB was required by law (and the certification process) to develop management plans that outlined strategies for environmentally appropriate use of all biological resources within their concessions. They were therefore motivated by external sources to build a constituency of support in favor of these efforts. In contrast, villagers do not yet see the negative consequences of over-hunting but profit from the short-term financial gain of these practices. Through its private TV and radio stations and political connections, CIB has strong capacity and power to influence people and thereby to build support for conservation activities. However, CIB lacks the technical knowledge (capacity) and internal motivation (although they possess strong external motivation). Although CIB superficially appears to be the most appropriate leading actor, the appropriate combination of actors is the partnering of CIB and WCS to ensure the success of this management role. Reprinted with permission from the Wildlife Conservation Society's Working Paper No. 28, 2006, pp. 75.

Roles of stakeholders evolve

Over time, the appropriate arrangement of actors and partners in a PSPC may change according to the social, political and environmental landscape together with the evolving attributes of the participating organizations. Assuming the partnership evolves in a positive direction for biodiversity conservation, the PSPC needs to be flexible enough to adapt to new circumstances and to engage new partners or to replace old ones. Growing pains are likely to occur within the individual organizations and the partnership as a whole. Since its inception, the BZP evolved considerably, as evidenced by the changing roles of each partner.

In its early days, the intensity of poaching required that BZP focus on law enforcement and awareness-raising activities. At the project inception, logging roads were lined with hunting camps, poaching of endangered species was rampant and people were almost completely ignorant of hunting regulations (Elkan & Elkan, 2005; Elkan *et al.*, 2006). The BZP responded to the immediate threats out of necessity before it moved on to address longer-term threats. As law enforcement and awareness-raising reduced the threat to endangered species such as elephants and apes, the BZP started the next phase of integrating local communities more directly into wildlife management.

Another example of the evolution of roles was the growth of CIB's involvement in the BZP as it sought FSC certification of its timber concessions (Boxes 2.5 and 2.6). In seeking FSC certification, CIB committed itself to high social, industrial and environmental standards. Principle #6 of FSC's International Standard defined the criteria for minimizing the negative impact of logging on the environment somewhat vaguely. For wildlife management, the standard included the protection of rare, threatened and endangered species and the establishment of conservation zones and protected areas. This standard and others are evaluated by independent auditors who decide whether the logging company has met the criteria for certification and, once obtained, whether the company is in compliance and can retain its certification.

Box 2.5 **The commitment to responsible forestry by the CIB (Jean-Michel Pierre, Dalhoff Larsen & Horneman A/S or DLH Group)**

With the declaration of the new Congolese forestry code in 2000, the CIB made a commitment to the sustainable management of the concessions granted to it by the government of Congo. CIB was the first company in Congo to make this commitment. Aware of the coming market requirements for tropical timber and the company's own environmental and social responsibilities as a major player in the sector, in 1999 the CIB started working towards the certification of its forest management and timber products. This process started with obtaining the label Keurhout for the Dutch market from 1999 to 2002. To differentiate itself on the European and North American markets and from competition from numerous illegal operations in the Congo Basin, the CIB developed its own label in 2003: the green leaf symbolizes its commitment to environmental and social responsibility. To establish a credible and sustainable strategy for responsible forest management, CIB began the process of obtaining FCS certification in 2004, the most demanding certificate and the only label to be supported by advocacy groups and NGOs involved in forest conservation.

The FSC label is based on the concept of sustainable development as defined at the Earth Summit in Rio in June 1992, which implies a forest management that is ecologically appropriate, socially beneficial and economically viable. FSC certification requires compliance with ten principles, each accompanied by specific criteria, which apply to all the forests and woodlands in the world. Forest management that meets FSC standards is not limited to the management of commercial timber, but includes management for all the biological, social and economic functions of forest ecosystems.

CIB received its first FSC certificate for the Kabo concession in 2006 and a second certification for the Pokola concession in May 2008. With the certification of the Loundoungou-Toukoulaka concession in February 2011, CIB achieved its goal of certifying all of it concessions and timber production over nearly 1.4 million ha. Moreover, the CIB's entire production is certified for its legality and traceability over its four concessions through independent audits conducted by SGS (Timber Legality & Traceability Verification and Controlled Woods).

Box 2.6 **Overview of the FSC principles and criteria**

The following ten FSC principles and their associated criteria describe how forests have to be managed (http://www.fsc.org/pc.html?& L=0). The principles and criteria can only be changed by a majority vote of the General Assembly of FSC members. All other FSC rules specify the requirements laid out in the FSC Principles and Criteria.

Principle 1. Compliance with all applicable laws and international treaties.

Principle 2. Demonstrated and uncontested, clearly defined, long-term land tenure and use rights.

Principle 3. Recognition and respect of indigenous peoples' rights.

Principle 4. Maintenance or enhancement of long-term social and economic well-being of forest workers and local communities and respect of worker's rights in compliance with International Labor Organization (ILO) conventions.

Principle 5. Equitable use and sharing of benefits derived from the forest.

Principle 6. Reduction of environmental impact of logging activities and maintenance of the ecological functions and integrity of the forest.

Principle 7. Appropriate and continuously updated management plan.

Principle 8. Appropriate monitoring and assessment activities to assess the condition of the forest, management activities and their social and environmental impacts.

Principle 9. Maintenance of high-conservation-value forests (HCVFs) defined as environmental and social values that are considered to be of outstanding significance or critical importance.

Principle 10. In addition to compliance with all of the above, plantations must contribute to reduce the pressures on and promote the restoration and conservation of natural forests.

By gaining FSC certification for the Kabo and Pokola concessions, wildlife and biodiversity conservation took on a financial interest for CIB. If the BZP's wildlife management program in these concessions had been ineffective, CIB would have been held responsible and it could have lost its certification, access to markets and profits. As a result, as CIB entered into the certification process, it became more involved in law enforcement and the protection of the rights of indigenous peoples. Moreover, CIB needed ecological and socio-economic data for its management plans and started to make demands upon BZP to produce the required data. The partnership evolved to fit the new circumstances.

Like CIB, the role of MEF also evolved over time. Biodiversity conservation in forest concessions was a fairly novel concept to the government of Congo in the late 1990s; it had just emerged from civil war and was focused on development using logging revenues. With growing awareness of the importance of forests to biodiversity conservation and carbon storage, responsible forest management took on new importance to the government. As the Ministry became more experienced in the issues, it took on bigger roles in forest concession management and in BZP, broadening its focus from an initial emphasis on law enforcement.

Financial contributions: Who foots the bill and what is the cost?

Capacity to implement a project or to conserve biodiversity has two attributes: (1) knowledge of what actions to take and the skills to undertake them; and (2) the financial, material and human resources to implement these actions. The technical capacity of each organization is discussed above; here we examine the capacity of each organization to mobilize the necessary financial resources to affect conservation.

Due to the spatial scope of its activities and the intensity of the threat to wildlife and biodiversity, BZP required a large investment of resources over the years. The annual operational budget of BZP averaged $825,000 between 2001 and 2006 (Table 2.2). This level of budget supported: (1) the construction, maintenance and operation of the BZP headquarters and equipment such as vehicles and boats; (2) salaries of the BZP staff; (3) equipment and operations for research and monitoring activities; (4) equipment and operations for the alternative activities program; (5) materials and operations

Table 2.2   **Approximate annual budgets (US$) for BZP from 2001 to 2006.**

	2001	2002	2003	2004	2005	2006
MEF[a]	50,600	50,600	50,600	50,600	50,600	50,600
CIB[b]	120,000	120,000	120,000	120,000	120,000	120,000
WCS[c]	831,773	725,099	728,023	667,085	671,859	606,349
Total	952,423	845,749	848,673	787,735	792,509	726,999

[a]Based on $3800/mo in salaries and $5000 in equipment each year
[b]Only the monthly contribution of $10,000 to law enforcement and excludes in-kind support.
[c]Based on a projection of project spending and not the actual numbers

for the environmental education and outreach program; and (6) equipment and missions for law enforcement. These activities are described in detail in Chapter 4. These figures do not include the actual government salaries of MEF personnel or infrastructure and logistical contributions by CIB, however. It also does not include investments by CIB in RIL, alternative activities or employee education relative to company hunting rules.

The level of activities grew with the level of investment. From 1999 to 2001, the project budget increased from $300,000 to $500,000 to $800,000, permitting expansion of activities to 200,000, 400,000 and 600,000 ha (Elkan & Elkan, 2005). During the first three years, annual costs varied from $1.25 to $1.50 per ha, with conservation activities focused on just the Kabo and Pokola concessions. Pokola required greater levels of financial investment in project activities because of its greater human population and hunting pressure (Elkan & Elkan, 2005). From 2001 to 2006, the average cost over the whole area of the landscape (Kabo, Pokola and Loundoungou concessions) was approximately $0.66 per hectare. According to Elkan & Elkan (2005), after costs of the establishment of the BZP infrastructure, law enforcement costs were highest (40–50% of the budget) followed by research and monitoring (17–19%). About 10% of the budget was dedicated to alternative activities, complementing CIB investment in this area, with another 10% funding environmental education and awareness-raising activities.

The MEF contribution was primarily though the appointment of personnel to the project. These personnel all received government salaries and included the MEF Coordinator, Brigade Chief, 2–3 Patrol Leaders and an Alternative Activity Technician. In most years, the MEF Coordinator also received a small operating allowance (~$1000) and/or equipment. Some of the equipment included a vehicle, motorcycle, computer and copy machine. In addition, the government supplied weapons and ammunition to the eco-guard unit.

Through international donors, WCS raised and contributed approximately 85% of the funding for project activities each year (Table 2.2). Donors included WCS private funds, foundations and individuals and governmental agencies. Funds leveraged for the project contributed to the salaries of personnel (WCS and MEF), equipment and materials, infrastructure, travel and project activities.

The financial contribution by the CIB was primarily directed at supporting the eco-guard unit. CIB paid $10,000 per month for the salaries of 25 eco-guards and field missions. CIB also supplied two transport vehicles to the project, monthly donations of fuel to run the vehicles and housing for approximately 60 project employees and their families in Kabo village.

Elements of a successful private-sector partnership for conservation

The BZP is one of the few long-term examples of private-sector partnerships for biodiversity conservation in the tropics with extractive industries. (Other types of partnerships include campfire programs with safari companies and ecotourism companies in parks.) In comparison to such other partnerships in Central Africa, the BZP is remarkable for its intensive, multi-pronged approach to wildlife management (Hurst, 2007). The success of this approach would not have been possible without the strong commitment of all three partners to responsible forestry, conservation of biodiversity and rural development. Despite its success, the road was not always smooth. Not only did the project face tough international criticism, at times the very partnership risked falling apart.

In addition to having the appropriate blend of partners, several other features of a PSCP contribute to its ability to function and to affect conservation. These features may vary depending on the scale of the threat to conservation, with large-scale threats requiring more formal management systems. Given its size, the management structure and functioning of the BZP was very formal and it grew more formal over time. Some of the key features of a PSPC that can make or break a partnership are discussed in the following.

Protocol of partnership

A protocol of partnership is essential for formalizing the relationship between partnering institutions and defining the roles and responsibilities of the partners. Formally, the BZP is a collaboration between the MEF, CIB and WCS (PROGEPP, 1999). It was created in 1999 through an official protocol signed by each of the partners that specified three goals: (1) environmentally appropriate management of the ecosystems in the forestry concessions adjacent to Nouabalé-Ndoki National Park; (2) protection of the park from the negative impacts of logging by managing wildlife in adjacent forestry concessions; and (3) collaboration with local people for the long-term management of their territories and natural resources.

The protocol accomplished two things. First, it defined the goals of the collaboration and the roles of each of the partners, including the financial, logistical and human resources that each organization was expected to contribute and the activities they were supposed to undertake. Second, it defined the management structure of the project, including the duties of key personnel.

The original protocol from 1999 was updated in 2008, with the key components staying the same (PROGEPP, 2008). Among other things, the new protocol established a steering committee and an evaluation committee to annually assess project activities and achievements and to provide guidance to the project managers. These committees consisted of members from each of the partner organizations, donors and outside experts.

Clearly defined organizational structure

When organizations are working together, it is essential that the organizational structure be defined clearly in order to avoid overlap, prevent conflict and permit the project to run efficiently. The organizational structure of the BZP was outlined in the protocol. At the project level, MEF and WCS co-managed the BZP and jointly oversaw project activities and staff, with each organization implementing the activities for which it was responsible. WCS also managed the budget, guaranteeing financial transparency for MEF and CIB. CIB played an indirect role in the management of the project; it assigned a person to liaise with the project on a day-to-day basis and contributed information and ideas in technical and steering committee meetings. The separation of CIB from the direct management of the project was chiefly a result of the project's law enforcement activities. Law enforcement was largely focused on logging company employees and vehicles, and was therefore a conflict of interest for CIB. This restriction neither lessened the role of CIB in the partnership, nor the impact of its activities towards improving the management of the logging concessions for natural resources.

The BZP was co-managed by personnel appointed by MEF and WCS. The MEF Coordinator and the WCS Principal Technical Advisor (PTA) planned and implemented project activities together[7]. The Coordinator was responsible for law enforcement and the PTA was responsible for donor-funded activities. A WCS-appointed Project Administrator managed the accounting and logistics. The MEF Brigade Chief managed the eco-guard unit. The MEF also appointed patrol leaders (usually 1–3) who led law enforcement missions in the field. WCS hired most of the rest of the staff according to the needs of the project, with the exception of two CIB drivers provided to the BZP for the transport of eco-guards.

The BZP was well developed in terms of infrastructure and personnel. By 2001 the BZP headquarters included its own office and base, personnel and administration. The base was located in the logging town of Kabo in the Sangha region of Congo. It consisted of approximately nine buildings including offices, warehouses, storehouses and a guesthouse paid for through donors to WCS; the International Tropical Timber Organization was one of the major contributors to office construction. Although the number of personnel varied slightly from year to year with fluctuations in the annual budget and activity plan, the staff generally consisted of at least 75 employees including 2–3 managers, 2–3 administrators, 25–40 law enforcement agents, 11 research staff, 4 alternative activity technicians and 10–20 drivers, janitors, cooks and laborers.

---

[7] With the signing of the 2008 protocol, the title of the WCS manager for the BZP was changed from Project Director to Principal Technical Advisor. This change was made in all the WCS-MEF projects in Congo and reflects the greater willingness and capacity of the Congo government to take a larger role in the daily management of their protected areas and conservation projects. Throughout the book we use the term Technical Advisor to be consistent with the current BZP management titles and protocols.

Opportunities for communication and collaboration

With the extensive array of activities performed by BZP partners (see Chapter 4), communication was essential for coordinating activities and sharing information. From the beginning of the project, the partners coordinated activities and goals through formal monthly meetings. As mentioned above, strong personal relationships between the CIB, MEF and WCS managers facilitated communication, but formal types of communication were particularly important for broadcasting information more widely through the different organizations. Some forms of formal communication that worked for the BZP included: (1) monthly technical meetings; (2) issue-based meetings scheduled when necessary; (3) bi-annual activity reports written by BZP staff and distributed to partners and donors; and (4) systematic exchange of documents and reports.

Conflict resolution

No partnerships are conflict-free. Government, industry and conservation organizations have different goals and, by their very nature, will likely disagree on issues and management approaches. An effective PSPC will design measures to resolve conflict, such as an advisory board or a steering committee. The BZP overcame several major conflicts over the years (Box 2.7). These ranged from discussions over how best to manage timber and wildlife resources, to arguments over the responsibilities of each organization to personal conflicts between personnel from the partner organizations. Programmatic disputes were typically resolved during annual steering committee meetings. These meetings were important for determining the larger vision of the project and setting the tone for project managers to work in the field. Serious personnel conflicts or charged issues such as the establishment of a sawmill sometimes required intervention from higher levels within each partner organization. When all other solutions were exhausted, the leaders of the PSPC organizations (WCS's Vice President of International Programs, CIB's Vice President and the MEF Minister) stepped in to resolve conflicts; they could implement policy decisions removed from the day-to-day stress of management on the ground.

Box 2.7 **Points of conflict between BZP partners**

- Construction of a CIB sawmill in the Loundoungou concession;
- construction of roads and logging camps in close proximity to the border of the NNNP;
- establishment of logging and wildlife management norms for management plans with 30-year timeframes;
- responsibility for management of BZP funds;
- role of the WCS and CIB in the management of the eco-guard unit when problems of corruption or poor performance arise;
- relative investment of BZP funds and effort into rural development versus law enforcement or conservation research; and
- personal conflicts among WCS, CIB and MEF personnel, stemming from the notion that one individual is not fulfilling his duties or is not sufficiently incorporating other partners in decision-making.

Definition of the roles of stakeholders

Given the different goals, objectives and priorities that often emerge in a PSPC, one of the most straightforward ways of preventing and mitigating conflict among partners is to ensure that the roles, rights and responsibilities of all partners are well-defined by formal protocols. The definition of roles assigns responsibility for management to the appropriate stakeholder and prevents overlap or doubling of efforts by different organizations. In the case of wildlife management, for example, eco-guards might be employed to enforce hunting laws. It must be clear who manages them and is responsible for their actions, failures and successes. When their work is deficient (e.g., poor performance or corruption), the responsible partner can immediately take action. This protects other members of the PSPC from blame for their failure. However, it is important to emphasize that by joining a formal partnership, the partner organizations are linked to each other. This has two ramifications: (1) the failure of one partner organization is likely to stain the reputation of the others (even if they are not directly responsible for the failure, they are complicit through the partnership); and (2) partner organizations usually agree to try to resolve their conflicts internally. Therefore, organizations should be willing to support each other and find constructive solutions to problems, rather than pointing fingers or assigning blame. The lack of defined responsibilities can expose organizations to lawsuits and loss of reputation, and can result in conflict among partners.

For natural resource management, some of the responsibilities that must be clearly assigned to a partner include: (1) assurance of food security of concession workers and local people; (2) collection and management of the biological and socio-economic data necessary to make management decisions; (3) incorporation of local communities into resource management activities; (4) management of different forest resources including wildlife, timber, non-timber forest products (NTFPs), fisheries, etc.; (5) conflict resolution among partners and other stakeholders; and (6) the degree of financial investment in wildlife and natural resource management.

The definition of roles and responsibilities should also include an explicit recognition of the actors to be consulted and the process of consultation for an industrial or management activity that could impact the environment. As an example, road construction to gain access to timber or mineral resources may fragment important habitat for endangered species or destroy sites sacred to indigenous peoples. It should be clear which partners or stakeholders are involved in road planning and how their input will be taken into consideration. What is the company's responsibility to inform versus consult with other stakeholders? Is the company beholden to the NGO partner or local communities if they object to company decisions, or can it continue with its plans despite their objections? The list of actors to be consulted should be based on criteria such as the proximity of people to an activity and their livelihood interests.

Getting your hands dirty

Although there are many ways that PSPCs could work, one of the strengths of the BZP model was that each partner played a role in implementing conservation on the ground. The MEF agents worked side-by-side with WCS managers and employees, and both were in frequent contact with CIB managers. Whether logging roads were being planned, poaching rings were being shut down or policy was being developed to protect indigenous rights, all three partners met and talked about the issues in order to come to solutions. It is critical that all partners 'get their hands dirty'. Working at a site alongside partners builds trust. On different occasions, both MEF and CIB expressed their appreciation for organizations such as WCS, who commit long-term to working in an area.

In fact, Henri Djombo (Minister of MEF) openly criticized advocacy NGOs such as Greenpeace that were perceived to parachute in, create a great deal of conflict around an issue and leave without helping to resolve problems. This criticism was made on the heels of Greenpeace's efforts to cause the government to reverse its decision that CIB build the Loundoungou sawmill within 16 km of the Nouabalé-Ndoki National Park[8]. Similarly, CIB compared on-site NGOs to those who visited once or twice a year, "left a 30-page report with recommendations" and did not assist in implementation.

A CIB manager articulated the importance of working together: "Most importantly, WCS was not only willing to conduct studies and make recommendations, but also to implement those suggestions. As a private company we are very good at throwing money at a problem. If we don't have the experience we can always just ship in an army of consultants. But once a consultant has made their report they leave again and then the company is left to try and implement it, which can be quite daunting if you do not master the subject."

The same can be said for government and industry. Because MEF and CIB were intimately involved in wildlife protection and resource management, they understood the challenges that make progress in conservation slow.

Hard times build trust

By 2010, the BZP partners had confronted and overcome many dilemmas and difficult issues. Working through problems and crises together builds an understanding of the issues that confront your partners and improves trust among organizations. An example of how hard times build trust among partners occurred at the creation of the BZP, when individuals and environmental groups criticized WCS for 'selling out' by working with CIB. In an online article, the World Rainforest Movement (2003) wrote:

---

[8] The Congo Forestry Code stated that a sawmill must be constructed in each concession, and MEF insisted that the law be followed. WCS argued that the sawmill should be moved farther from the park and warned that the installation of a permanent logging town and sawmill would increase hunting pressure in and around the Nouabalé-Ndoki National Park. CIB argued against the construction of a new sawmill claiming that it was costly and unnecessary, but the company never proposed a different location for its construction.

"The WCS has long known of CIB's impact on wildlife and its involvement in the extraction of bushmeat but has done little to give these findings prominence. In 1995, the WCS and a team of IUCN assessors even co-signed a Protocol with CIB that repudiated 'unjustified attacks' made on CIB – the evidence in the video documentaries. CIB, which has been unwilling to submit its forestry operation to scrutiny by independent certification processes like FSC, has been able to vaunt its close relations with WCS to fend off criticism of its operations: 'I have opened my concession for research . . . for forestry and wildlife studies', claims CIB owner Hinrich Stoll, my company is 'working very closely with the Congolese National Park, Nouabalé-Ndoki, which is managed by Mr JM Fay of the Wildlife Conservation Society (WCS), the oldest non-governmental ecological organization in the world'."

These allegations are an example of the risks the partners took by entering into a PSPC. WCS risked tarnishing its reputation by working with a logging company, while CIB was attacked for its shortcomings despite being progressive in its integration of wildlife management into its operating procedures. By 2003, CIB welcomed FSC into its concessions to gain certification of its Kabo concession in 2006.

Weathering criticism reinforced the working relationship between CIB and WCS by building respect and appreciation for the risks taken by the partner organizations. A CIB manager explained it this way: ". . . CIB's initial cynicism of WCS's intentions got replaced by a certain amount of good faith, especially the periods where the two organizations had to stand together and be supportive of each other due to outside criticisms/attacks. A classic case of: 'enemies of my enemies are my friends'. Of course that only laid the basis, and I think in time CIB's emerging awareness of certification, and what it would mean for our business, became the driving factor [of its commitment to biodiversity conservation]. No cost or burden was too much to ask as long as CIB saw it as an investment that would ultimately lead to increased returns."

The trust built through overcoming hard times can only be capitalized upon if there is an institutional memory of shared history. Institutional memory can be lost quickly if there is no continuity in the individuals that work within the PSPC.

Conclusion

This chapter has discussed some of the components of an effective PSPC. The relationship between a conservation organization and a company can be intricate and complicated. First, both partners do not necessarily dance to the same music. While the conservation organization waltzes to the symphony of big ideals and far-off goals, the company marches to the profit-focused cadence of its shareholders and clients. In many cases, a third partner (government) must also keep step. Sometimes it is not clear who is leading the dance; however, what matters the most is that you are dancing.

3

Land-use Planning in a Co-management Context: Establishing Access Regulations that Promote Biodiversity Conservation and Support Local Livelihoods

Connie J. Clark[1], John R. Poulsen[1], Germain A. Mavah[2], Antoine Moukassa[4†], Dominique Nsosso[5], Kibino Kimbembe[4] and Paul W. Elkan[3]

[1]Nicholas School of the Environment, Duke University, Durham, NC
[2]School of Natural Resources and Environment, University of Florida, Gainesville, FL and WCS Congo Program Wildlife Conservation Society, Brazzaville, Republic of Congo
[3]WCS Africa Program, International Programs, Wildlife Conservation Society, Bronx, NY
[4]Wildlife Conservation Society, Brazzaville, Republic of Congo
[5]Direction de la Faune et des Aires Protégées, Brazzaville, Republic of Congo

The strict protection of individual species and their habitats was once perceived as the sole means to conserve wildlife. This approach led to the creation of parks and reserves that now cover approximately 12% of the global land surface and approximately 8% of the remaining tropical forests. Parks and reserves

†deceased

Tropical Forest Conservation and Industry Partnership: An Experience from the Congo Basin, First Edition. Edited by Connie J. Clark and John R. Poulsen.
© 2012 Wildlife Conservation Society. Published 2012 by John Wiley & Sons, Ltd.

generally limit access to resources, sometimes banning all human activity, and prioritize biodiversity conservation, ecological integrity and wilderness preservation (Soule & Terborgh, 1999; Holt, 2005; Hutton *et al.*, 2005). Although integral to conservation efforts, critics of protected areas argue that they ignore the needs and rights of rural people by placing large tracts of forests vital to the livelihoods of local populations completely out of bounds (Hackel, 1999). Because this is often contrary to the interests of adjacent communities, hostility between rural people and protected area managers can ultimately compromise the success of protected areas (Ghimire & Pimbert, 1997).

The recognition that rural people have both a right to natural resources and a motivation to conserve them led to the rise of community-based conservation as a popular model for resource management (Gibson *et al.*, 2000; Agrawal & Ostrom, 2001). Community-based management (CBM) approaches gener- ally assume, rightly or not, that because the livelihoods of local people are tightly connected to natural resources, local communities have both better knowledge of and higher stakes in effective resource management than does the state (Tuxill & Nabhan, 2001; Wilshusen *et al.*, 2003). The varied results of CBM projects challenge the accuracy of this assumption and suggest that complete control of resources by local actors is not always the best strategy for sustainability. In many cases, access regulations dictated by local communities have been insufficient to sustain the resource base. In others, the scope of man- agement problems are beyond the capacity of local institutions (Armitage & Johnson, 2006; Berkes, 2006); local institutions can be vulnerable to pressures and incentives that originate at other levels of social, political or economic organization (e.g., illegal logging operations and well-organized poaching networks). The appropriateness of community-based approaches are there- fore site-specific, depend on particular characteristics of the resource and are not a magic bullet for sustainable resource management and biodiversity conservation (Castillo *et al.*, 2006).

The emergence of conservation strategies that embrace a landscape man- agement paradigm has facilitated a less dichotomous approach to natural resource management and biodiversity conservation. Landscape-scale con- servation generally extends into a geographic space beyond protected areas or community reserves: a space inhabited by multiple actors with various interests and often poorly defined, overlapping and contested access rights. Conservation success in complex landscapes necessitates management systems that distribute authority across institutions, rather than concentrating it in just one institution (e.g., the state versus local communities, Barrett *et al.*, 2001). We refer to such systems as *collaborative management* (CM) systems, defined by the World Conservation Congress as a partnership in which "government agencies, local communities and resource users, nongovernmental organizations and other stakeholders negotiate, as appropriate to each context, the authority and responsibility for the management of a specific area or set of resources" (IUCN, 1996). The idea is that an agency with jurisdiction over an area (usually a state agency) might develop "a partnership with other relevant stakeholders (primarily including local residents and resource users) which specifies and guarantees their respective functions, rights and responsibilities with regard to [the area]" (Borrini-Feyerabend, 1996).

Advocates of the collaborative management approach argue that community-based conservation should be conceived as one component of a multi-faceted management system that also involves support from complementary institutions with the capacity or mandate to better deal with governance issues, multiple stakeholder objectives and the creation of problem-solving networks to address the complexities of resource management in a globalized world (Kooiman, 2003; Carlsson & Berkes, 2005).

The ecosystem management strategy adopted by the Buffer Zone Project (BZP) is an example of a collaborative management approach between the Congolese Ministry of Forest Economy (MEF), Congolaise Industrielle des Bois (CIB), the Wildlife Conservation Society (WCS) and the local communities. The management system engaged the capacity, mandate and views of each partner to strike a balance between goals that were initially perceived to be in conflict (as discussed in Chapter 2). Goals that may be perceived to be in conflict include conserving biodiversity, promoting economic development though timber exploitation and maintaining local livelihoods and cultural values through the integration of local populations into the decision-making process.

Among the most important aspects of this management system was the development of a zoning system and access regulations to help foster the sustainable harvest of the floral and faunal resources[1]. The BZP zoning process included: (1) definition of the management goals and priorities of the project

---

[1] Conservation requires that norms be developed for regulating access to and metering use of natural resources. The development of a BZP zoning plan was a means of bringing together multiple actors in a co-management context to define who has legitimate access rights to what natural resource and how much of each resource legitimate resource-users should be allowed to harvest in a given time period.

partners; (2) examination of the social and ecological context of the area to be managed; (3) initiation of a participatory research program to describe and acknowledge traditional patterns of resource use, access rights and regulations; and (4) agreement on and official adoption of access regulations to mitigate threats to the resource base[2].

The BZP partnership acknowledged that fair management regulations must be based, when possible, on the traditional access rights and land-use practices of local communities. In response, the BZP developed and implemented a participatory research program to evaluate current and past patterns of local resource use. Specifically, the program sought to understand how and where local forest-dependent populations use forest resources and determine to what degree these patterns of resource use are sustainable[3].

The goal of this chapter is to describe the participatory research program initiated by the BZP partnership to develop access regulations that reflect the needs, concerns and traditional practices of local communities. Boxes 3.1 and 3.2 describe a complimentary participatory program introduced by CIB and The Forest Trust (TFT) in the CIB concessions.

Box 3.1 **Participatory mapping of forest resources (Scott Poynton, Executive Director, The Forest Trust)**

Under the BZP partnership, in the late 1990s CIB embarked on a path to improving its understanding and management of the ecosystem from which it was harvesting timber. Over time, CIB went even further and developed procedures that sought to minimize the inevitable impact of its timber extraction on both the forest and the indigenous communities that depend upon it: around 17,000 individuals of whom 9000 are semi-nomadic forest-dwelling peoples. Historically, most indigenous communities in the Congo Basin have been given limited opportunity

---

[2] Spatial delineation of resource-use zones determined where and to whom access rights to forest resources were allocated. Within each zone, Congolese wildlife and forestry laws dictated access regulations regarding how much and when resources can be harvested.

[3] Where resource use is sustainable, participatory research coupled with a co-management structure can help secure local rights against possible losses caused by logging operations. If resource use is unsustainable, the process can help identify strategies to prevent or reverse the degradation of natural resources.

to participate in land-use decisions directly affecting the resources upon which they depend. CIB faced the challenge of locating and communicating openly with a semi-nomadic non-literate people of a totally different linguistic and cultural background. This raised the question of how best to establish a viable mechanism to ensure an effective and fair co-management of the forest resource.

As part of the process, CIB committed itself in 2004 to achieving Forest Stewardship Council (FSC) recognition for its environmental and socially responsible forest practices. CIB management knew that it needed help to better understand and respond to indigenous community concerns, so it engaged with The Forest Trust (TFT) to implement a resource and cultural mapping program which later became known as the Indigenous Peoples Voices project. Financing was secured from the World Bank Development Marketplace and expert advice and participation was sought from Dr Jerome Lewis (an anthropologist who lived on-and-off for a decade within these communities), John Nelson (Africa Policy Advisor to the Forest Peoples Programme) and technology partner Helveta. After an extensive and inclusive consultation process with the communities and other stakeholders, a unique icon-driven GPS mapping process was created. The process differed from traditional mapping exercises in use at the time in two significant ways: it was designed with community input (both men and women were actively consulted) and it did not require literacy skills.

The community-produced and -owned maps identified locations of cultural value and resource importance, and formed the basis of a communication platform from which community involvement in forest management decision-making was sought. After a formal negotiation process to establish consensus on key resources (based upon the maps), CIB incorporated the data into its harvesting plans to ensure its timber operations do not affect the important resources. In addition to the electronic map recordings, the communities themselves physically marked these same resources in the forest. As a result, there is a deep understanding at CIB of the importance of these areas to the poorest communities in the region and, in turn, the communities feel their concerns were actually taken into consideration.

Where differences in social status and language often make direct communication ineffective, this mapping technology enabled both parties' interests to be considered equally. Most importantly, this was the first time such a win-win situation was achieved in the Congo Basin. As with any new initiative, major challenges were faced. Gaining the trust of communities that historically had been disenfranchised both politically and economically was not easy. This was however overcome with time through the recruitment and training of Mbendzélé community members and as they gained experience working with the project.

Based in part on the success of this activity, all of CIB's concessions totalling 1.3 million hectare are now FSC certified. This ground-breaking project will remain a model for others in the industry. The innovative use of technology has made possible new relationships between indigenous peoples and industrial-scale logging operations with enormous potential for replication throughout the entire region.

Box 3.2 **A voice in the forest on 88.0 FM (Scott Poynton, Executive Director, The Forest Trust)**

In addition to its mapping initiative (Box 3.1), The Forest Trust (TFT) created a second programme under its Indigenous Peoples Voices project: Biso na Biso (BnB), the first indigenous-language community radio station in the Congo Basin. Based in Pokola and hosted on site at CIB, BnB serves indigenous forest-dwelling semi-nomadic Mbendzélé and Ngombé and local Bantu populations living in the Sangha and Likouala districts of north Congo. Biso na Biso (meaning 'between us' in Lingala) broadcasts in twelve local languages including Lingala, the country's lingua franca. The idea behind a community radio in the forest was born from the need to raise awareness among local populations about regional forest management issues and to open and maintain a line of communication between CIB and the communities in its concessions.

BnB promotes diverse regional cultures and endangered languages, addresses the human rights issues inherent in responsible forest management and provides important local information relating to community services such as health care and education. It also strengthens relations between Mbendzélé and Bantus. Key objectives are to:

- serve as a platform for exchange and information among local people;
- facilitate dialogue between forest dwellers and industrial operators and other local stakeholders;
- value the twelve linguistic groups and cultural heritage of the Pokola region, including Mbendzélé and Bantu communities;
- provide opportunities for marginalized groups to access and use the medium of radio (women, young and old people) for communication;
- provide education and information on natural resource management in the rainforest; and
- inform about issues surrounding sustainable forest management, CIB's logging activities, the principles and criteria of FSC and Free, Prior and Informed Consent practices.

Over 1000 individual productions have been made since the launch of BnB in June 2009; these include magazines, interviews, stories and reports with several hundred Mbendzélé and Bantu songs recorded live in the forest or in the studio. There are varied regular broadcast themes from participative management of natural resources (fauna and flora), environment (water, forest, living environment), climate change, hunting, fishing, agriculture, community life, HIV/AIDS, health and hygiene, oral traditions, education and community initiatives, women's issues, civic life, FSC and CIB.

BnB has four permanent staff, five volunteer leaders and ten community reporters (all based in their respective communities). These volunteers and reporters are non-professionals, Mbendzélés and Bantus (men and women), trained since the start of the radio in practical community journalism. Advisory editorial and listener's committees meet regularly and advise staff on production and programming to develop creative, relevant and innovative content.

The BnB signal covers a 120 km radius into the 1.3 million hectares of forest under CIB's management and can be heard in several remote areas close to the Cameroonian border in the outskirts of Nouabalé-Ndoki National Park. It has a 1000-watt transmitter at its primary relay site 21 km from the broadcasting base in Pokola, powered by solar panels. Emissions are ensured by a full state-of-the-art digital production studio and extensive mobile recording equipment. Solar- and dynamo-powered radio receivers have been distributed throughout villages and camps. An additional benefit of the design (apart from no need for electrical power) is that it does not require batteries.

In a relatively short time BnB has become the area's new tool for communication, for cultural exchange and dialogue, for awareness of natural resource management and for promotion of traditional heritage. It has become the common platform that was previously lacking in the region. Through the recording of traditions, stories, rhymes, lullabies, songs, proverbs and games, BnB enhances the use of languages and cultures which might otherwise be threatened with extinction. In effect, it is helping is to balance modernity and traditional values while providing a platform to raise awareness of the need to respect biodiversity and human rights.

BnB is an innovative project, a medium for the peoples of the forest to communicate with each other and with outsiders. It is run by the very people it is designed to serve and addresses issues at the heart of sustainable development: the equitable management of natural and cultural resources. For the first time through the radio, these communities have a digital catalogue that ensures the preservation of their traditional collective memory, a forest culture that has been in place for centuries.

BnB is made possible thanks to a number of local partnerships with SAM (Sangha Medical Assistance), Coska (Commonwealth of Semi-nomadic Peoples in Kabo), ACOBAK (Bantu Community Association of Kabo) and the Foyer Fréderique (Pygmy Support Centre, Pokola). The World Bank's Development Marketplace provided start-up funding for the feasibility study and equipment purchase and CIB and Fondation Chirac support current running costs.

Participatory research program and land-use planning

Participatory research methods were designed to incorporate local knowledge, perspectives and priorities in the development of access regulations (Moukassa & Kimbembe, 2003; Moukassa *et al.*, 2005; PROGEPP, 2005a, 2005b). Participatory methods offer local people a role in research and planning that can result in locally appropriate solutions to resource management. The specific goal of the research program was to understand the social, economic and cultural factors that dictate land-use practices and ensure they are respected. To do this, the BZP worked to: (1) map traditional territories and resource uses of each ethnic group within the CIB concessions; (2) understand the traditional rules that governed resource use before the arrival of logging operations; and (3) evaluate the spatial occupation of the semi-nomadic Mbendzélé and Ngombé[4] communities to gain an understanding of their land-use patterns in space and time.

Step 1: Completion of participatory 'sketch maps'

To gain an understanding of the spatial and temporal patterns of resource use, we first engaged local communities in a variety of land-use planning processes including: mapping of traditional land and resources, mapping of historical patterns of land-use and mapping of sacred sites and trees (Figure 3.1).

Participatory mapping involves communities plotting information about their current and past zones of resource use. It empowers local communities and indigenous people to become more involved in resource management and environmental protection. Participatory mapping helps construct visible images that reflect the story any given local informant is telling and can help local populations depict resources and geographical features, graphically manifesting the significance they attach to them. Depending on the comfort of the local informant, maps can either be traced on paper or on the ground using local materials. This process sheds light on systems of traditional authority to regulate resource access and use. It also helps reveal information about

---

[4] Mbendzélé and Ngombé communities are forest-dwelling semi-nomadic ethnic minorities indigenous to northern Congo.

Figure 3.1 **A local woman paints a sacred tree to protect it from extraction or damage during logging operations as part of a CIB/TFT participatory community mapping project. Photo by The Forest Trust (TFT).**

conflicts, overlap among communities in zones of use and areas where rights and responsibilities are ambiguous. The mapping technique adopted for this project entailed the production of participatory sketch maps for each village.

The BZP social research team conducted participatory mapping exercises in all concession villages between 1999 and 2004 (Moukassa *et al.*, 2005; PROGEPP, 2005a). Men and women of both Bantu and Mbendzélé or Ngombé origin (initially in separate groups; see Box 3.3) were invited to participate in a series of village meetings (following prior approval by village authorities). The research team explained the participatory mapping process to the village and demonstrated examples of previous mapping exercises. Once the participants understood the process, they drew maps of their territories and natural resources on the open ground, using local materials such as stones, twigs and soil (Figure 3.2). To begin the process, the village nominated one or two people to 'hold the stick' (quite literally). The village representative then translated his or her perception of village resource use to the map on the ground. Other participants voiced their opinions of the map (sparking much discussion among community members) and the stick was passed onto other villagers to draw their views of resource use. In this way, the process continued

Figure 3.2   **A villager sketches the resource zones around his village during the participatory mapping process. Photo by Buffer Zone Project.**

until consensus was reached regarding the placement of: (1) landscape features that could orient the map geographically; (2) forest resources, particularly wild game, leaves and fruits and insects, building poles and fuel wood and water; and (3) sacred sites, trees or areas of cultural value. When villagers felt they had completed an accurate ground sketch map, researchers photographed the original map and then traced it onto paper with a pen. Village participants verified the traced version of the map and signed a statement that they were satisfied with the results of the process (Figure 3.3). The participatory mapping process was repeated with elders of the village to record historical patterns of land use and to gather information regarding how and why resource use changed over time (Figure 3.4).

Box 3.3   **Integrating marginalized communities in land-use planning**

The co-management process can expose disadvantaged or minority groups to manipulation and/or control by more powerful groups. The more advantaged individuals and groups in societies are likely to be most capable of exploiting participatory management systems. Taking special measures to support marginalized groups can counteract this.

In the CIB concessions, the Mbendzélé (a clan of the larger Aka peoples) were traditionally hunter-gatherers with semi-nomadic customs. Under the influence of the government settlement policy and the search for employment, many Aka communities have settled in villages located close to Bantu villages. This co-existence often disturbs the organization of the Aka society and modifies their customary relations with the Bantu. Once settled, Aka are often obliged to abide by Bantu social norms and are sometimes deprived of their traditional rights to land and natural resources. It is in such a context that the BZP was initiated. To achieve co-management, the BZP had to carefully navigate the social context of Bantu/Mbendzélé interactions and, in particular, make sure the Mbendzélé had a voice in the participatory mapping process. To this purpose, activities were designed to address the Bantu and Aka communities individually before bringing them together as a group. As a second step, BZP made opportunities for discussion between the Mbendzélé and Bantu communities.

Differences in privilege and power are quite common between social actors such as governmental agencies and local communities. Local communities, however, also possess inequalities based on caste, class, gender, ethnic origin, age groups, etc. Unfortunately, conservation initiatives can exacerbate such inequalities if they are not explicitly addressed.

Some ideas for facilitating the promotion of marginalized communities into participatory processes (adapted from Edmunds & Wollenberg, 2002), include:

- informing participants of the organization or group that the facilitators represent;
- giving disadvantaged groups the option of not participating in negotiations (in order to avoid being more 'visible' to powerful stakeholders);

- creating possibilities for disadvantaged groups to use alliances with more powerful groups in negotiations;
- acknowledging the right of disadvantaged groups to identify topics that they view as inappropriate for discussion (or 'non-negotiable') in the negotiations;
- acknowledging that groups may not wish to unconditionally support an agreement;
- assessing the likelihood that external events require revisions in agreements and making provisions for disadvantaged groups to be involved in those revisions; and
- approaching negotiations as one strategy among several that disadvantaged groups may pursue simultaneously, and helping them to identify alternative strategies in case the good will of other actors does not last.

Step 2: Transfer of sketch to GIS-based maps

To facilitate communication between villagers and other stakeholders and to assure geographical accuracy, hand-sketched maps were transposed to conventional topographic maps. To do so, the social team accompanied Bantu and Mbendzélé community members into the forest with the sketch map to register GPS locations for as many of the landscape features, resource areas and sacred sites as possible. GPS positions were recorded for current and past forest camps, ancient village sites, cemeteries, sacred sites, river courses, important trees and borders of traditional territories. Researchers then used GIS techniques to geo-reference the village map and transpose it onto topographic maps of the area. Upon completion of the geo-referenced map, researchers returned to each village with copies of both the initial sketch map and the transposed map for approval by the local population. This exercise served as the foundation for integration and recognition of traditional zones of use into the CIB company rules and management plans as 'village hunting zones' (Figure 3.5).

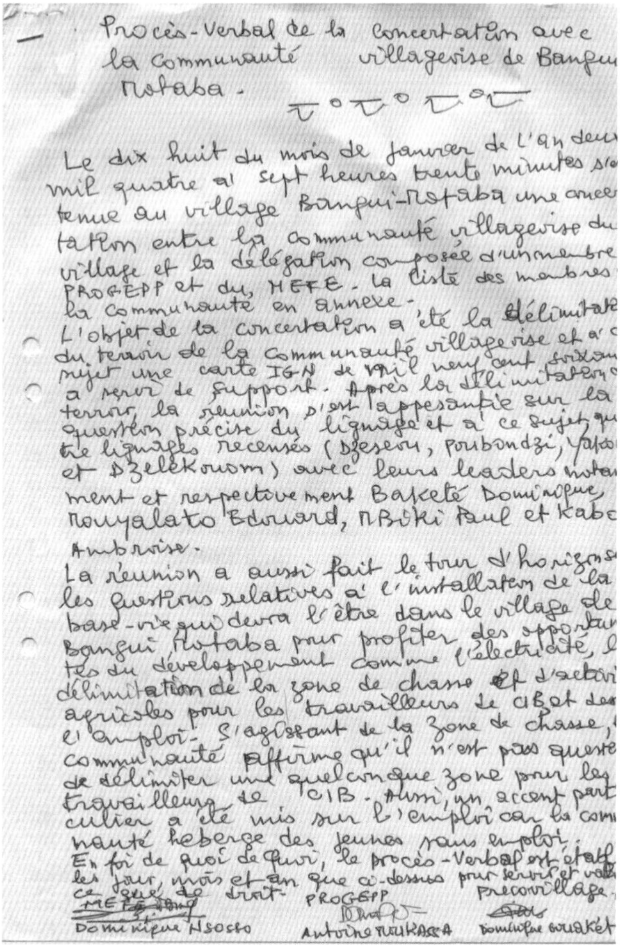
Procès-Verbal de la Concertation avec
la communauté villageoise de Banguí
Motaba.

Le dix huit du mois de janvier de l'an deux mil quatre à sept heures trente minutes s'est tenue au village Bangui-Motaba une concertation entre la communauté villageoise du village et la délégation composée d'un membre PROGEPP et du MEFE. La liste des membres de la communauté en annexe.
L'objet de la concertation a été la délimitation du terroir de la communauté villageoise et à ce sujet une carte IGN de mil neuf cent soixante a servi de support. Après la délimitation du terroir, la réunion s'est appesantie sur la question précise du lignage et à ce sujet quatre lignages recensés (Dzeseru, Poubondzi, Yakoo et Dzelekouom) avec leurs leaders notamment et respectivement Bakete Dominique, Nouyalato Edouard, Mboki Paul et Kabe Ambroise.
La réunion a aussi fait le tour d'horizon sur les questions relatives à l'installation de la base-vie qui devra l'être dans le village de Bangui Motaba pour profiter des opportunités du développement comme l'électricité, la délimitation de la zone de chasse et d'activités agricoles pour les travailleurs de CIB et des emplois. S'agissant de la zone de chasse, la communauté affirme qu'il n'est pas question de délimiter une quelconque zone pour les travailleurs de CIB. Ainsi, un accent particulier a été mis sur l'emploi car la communauté héberge des jeunes sans emploi.
En foi de quoi, le procès-verbal est établi les jour, mois et an que ci-dessus pour servir et valoir ce que de droit.

MEFE PROGEPP Precovillage

Dominique Nsosto Antoine Nzikassa Dominique Bouaket

Figure 3.3 **Example of a *procès verbal* summarizing the process of determining village use zones.**

Step 3: Integration of zoning system into company rules and concession management plans

As a result of the participatory mapping process, a zoning system was created early in the project that included village hunting and no-hunting areas.

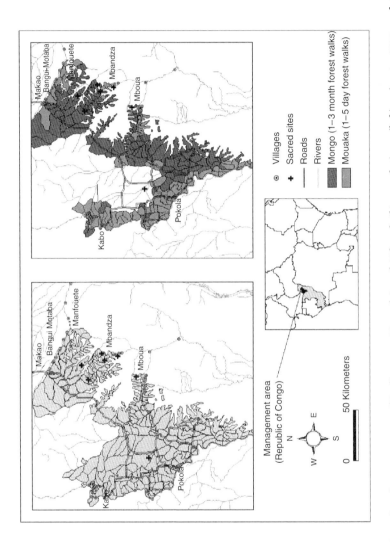

Figure 3.4 **Land-use patterns of indigenous Ngombé and Mbendzélé populations (left) prior to 1968 and (right) present day. (Source: Moukassa et al., 2005.)**

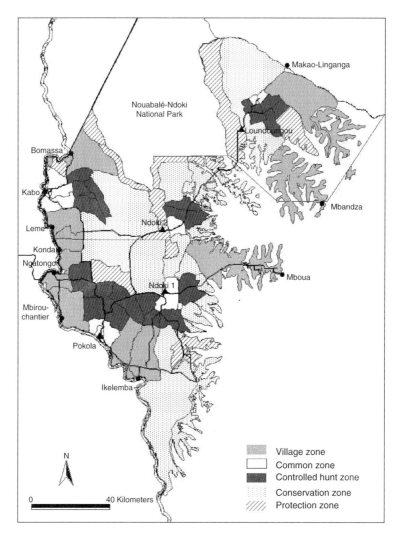

Figure 3.5   Map of CIB concessions with the hunting zones in the Kabo and Pokola concessions and proposed zones in the Loundoungou concession. Map by G. Mavah.

Several villages adopted specific regulations for their hunting zones, and even requested assistance from BZP to protect the zones from hunters from outside the community (Elkan *et al.*, 2005). The CIB company rules also included a provision specifying that its employees had to respect the zoning system. This zoning system became the basis for the village resource extraction zones that were later formally adopted in the concession level management plans.

The final management plan zoning system was determined through negotiation among stakeholders and adopted by consensus. Because only CIB (through management plans for the concessions) and the government possess the legal mandate to recognize a consensus zoning system, they led the negotiation process. Following several months of negotiation, a multi-layered zoning plan was proposed that integrated the development and extractive needs of the government and CIB, the biodiversity conservation concerns of WCS and the traditional territories of local communities (Congolaise Industrielle des Bois, 2006, 2008, 2011).

The final land-use plan had two structural levels. The first level defined where logging activities would occur. It attempted to balance the concerns of local populations and conservation with the goal of maximizing timber production and economic profit within the norms of sustainable timber production. Zone types are defined by the National Forestry Management Directives and include five land-use categories: (1) the production series is set aside for logging operations and economic production; (2) the conservation series guarantees the existence of timber species and protects biodiversity, wildlife and landscapes; (3) the protection series safeguards fragile habitats, particularly watersheds, watercourses, swamps and soils that could be degraded by erosion; (4) the community development series is reserved for use by local populations to harvest natural resources for their livelihoods and community development; and (5) the research series delimits areas that can be used for ecological and forestry research. In the Kabo concession, 72.3% (2140 km^2) of the area is included in the production series, 20% (59,300 ha) is included in the protection series, 5.1% (15,100 ha) in the conservation series and 2.6% (7600 ha) in the community development series. The entire area is included in the research series.

The second level of zoning involved the delineation of village resource extraction zones, particularly hunting, within the production and community development series. Village zones are subdivided into zones specifically for indigenous villagers, zones for both indigenous villagers and logging company employees and their families and zones for the controlled hunt (a bi-monthly hunt organized for CIB employees[5]). Conservation zones prohibit hunting with firearms, but allow hunting and trapping with traditional weapons fabricated from natural materials (e.g., spear, crossbow, woven nets, etc.). Fishing and gathering are allowed throughout the year. Protection zones conserve areas of particular importance for endangered species or biodiversity (e.g., the buffer around the park borders and large forest clearings) and all hunting, either modern or traditional, is prohibited.

The conservation and protection zones afforded protection to game populations and key habitat, and presumably served as a source of wild animals to replenish stocks in neighboring hunting zones. The Kabo concession, for example, was divided into village hunting zones (139,600 hectares, 47% of the concession), conservation zones (115,400 hectares, 39% of the concession) and protected zones (41,300 hectares, 14% of the concession). The zoning protects indigenous rights by allowing year-round hunting with traditional techniques by Bantu or Mbendzélé/Ngombé in both the village hunting zones and conservation zones (86% of the concession).

The village-use zones reflected the participatory mapping process as closely as possible. They specifically limited access to wildlife to hunters from the adjacent villages and exclude immigrant populations. To a large extent, the different zones took into account the different cultures of villagers and Mbendzélé peoples. Whereas villagers perceive the land around the village as territory belonging to the village and off-limits to others, Mbendzélé believe that the entire forest was created by *Komba* (God) for all to share (Lewis, 2002). By delineating village hunting zones (in which Mbendzélé are also permitted to hunt) and, at the same time, allowing traditional hunting and gathering throughout most of the concessions, the system addressed the traditional access systems and perceptions of land-use rights of both cultures.

---

[5] The creation of controlled hunting zones was an incentive negotiated with the workers union to encourage company workers to respect company hunting rules. Through the controlled hunt, workers were allowed bi-monthly hunts during the hunting season at sites far from villages (these workers also had the legal right to hunt in the 'village' zones with which they were connected). A group of workers was transported in a company vehicle to the controlled hunt zone, thereby privileging workers over non-workers, who were not allowed to take part in the hunt.

Step 4: Validation of the land-use plan and the legal recognition of indigenous rights

Upon completion of the zoning process each village was revisited by the social research team and a public meeting was held to discuss the proposed zoning principals (including wildlife protection and forestry exploitation) with community members. Importantly, each village was given a voice in the process and the traditional zones described in the participatory mapping exercises were retained for subsistence hunting and gathering activities in the final zoning plan. Finalized maps were validated in a public assembly with the participation of representatives from each community and the MEF, WCS and CIB.

The community mapping process was a critical step in the development of access regulations to guide resource use in the CIB concessions. Although imperfect, the process was a first step towards legally empowering local people to manage their own natural resources[6]. A key lesson from the process is that participatory mapping can be an effective method to promote land tenure. Even though the national government does not officially acknowledge traditional land tenure, the adoption of the management plan incorporated village views of 'tenure' into a government-adopted management plan. The consensus behind this map gives it legitimacy in political debates, if and when the government is open to such debate.

Step 5: Evolving toward a more equal co-management system

As a means of strengthening village institutions and formally incorporating indigenous populations into resource management, the BZP organized resource management committees of villagers and Mbendzélé or Ngombé in 2006. Resource management committees offer a conduit for information exchange with local communities and a structure for implicating people in the development of hunting rules and zones. In this way, BZP sought to

---

[6] In national law, people indigenous to an area of Congo and immigrants from other parts of Congo have equal rights to the use of natural resources. Adoption of a zoning system based on land-use practices of indigenous people was a positive step towards reinforcing local authority over their traditional hunting, fishing and gathering zones. However, progress must still be made in preventing management decisions in logging concessions from marginalizing indigenous populations.

empower communities to make decisions that concern wildlife management (e.g., developing hunting rotations around villages, reducing harvest of rare species or developing systems to restrict the use of hunting zones by outsiders, if necessary). The goal of these committees was to gradually shift many of the responsibilities and resource management activities conducted by other partners to the communities.

Role of management committees

Management committees are usually voluntary community organizations with a purpose of involving communities through participation in common action (Florin & Wandersman, 1990). These committees can play many unique roles that draw upon their strengths and capacities as community-oriented institutions. The management of common property resources can act as a powerful catalyst for communal institutions such as resource management committees. In fact, the long-term goal of BZP was to help manage common property while balancing individual and collective interests. Through resource management committees, communities can, over time, gain experience and take action to manage wildlife sustainably and improve their livelihoods. We facilitated the establishment of resource management committees through a three-step methodology depicted in Figure 3.6: (1) preparation of partnerships; (2) negotiation of management organizations; and (3) implementation and monitoring.

Preparation of partnerships

The BZP first defined a protocol for the establishment of the village-based resource management committees. Because of the interactions between macro- (national) and micro- (village) levels of governance and the existence of political elites (Gami & Doumenge, 2001), the project communicated with all the different political levels to get buy-in for the establishment of the committees. Socio-economic teams notified the Congressional representative of the area by mail; met with regional and district authorities, including forestry officers; notified village committees by mail; and planned visits to villages to discuss the creation of the committees.

In traditional villages, people in possession of knowledge of the customary ownership of land in and around the village were identified. This was necessary

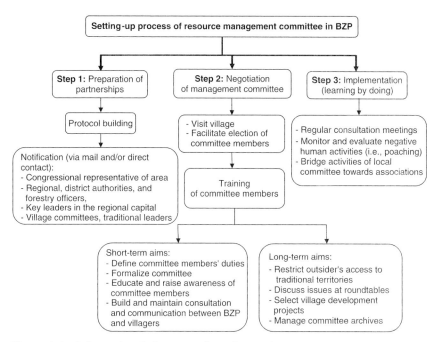

Figure 3.6 **Schematic of the procedure for setting up resource management committees.**

to acquire detailed information from multiple sources about traditional leadership and land ownership in each village according to customary law. (Customary law was treated as an informal tool because, though still relevant for our purposes, it was formally abolished in 1983).

Negotiation of management organizations

Local communities were composed primarily of two ethnic groups; Bantu and Mbendzélé. Both groups were incorporated into the resource committees so that they were equally represented and to build social trust between the ethnic groups. Villages elected committee members, including a committee president, vice president, two secretaries, one treasurer and two representatives. Once the committee members were chosen, BZP organized workshops to build the management capacity of the committees. The short-term aims of the workshops were to: specify the duties of committee members; develop a constitution to formalize the committees; raise awareness of the environmental impacts of extractive activities; and build a relationship of consultation and communication between BZP and the committees.

Implementation and monitoring

The objective of the resource management committees was to manage forest resources, implement strategies, evaluate their impact and refine them. Having built strong relationships with the committees, the BZP continued to consult with the committees and to evaluate their progress and efforts in 2010. At the same time, BZP offered direct assistance in helping control access to traditional territories by enforcing wildlife laws when the committees informed the BZP of poaching. The management roles that local communities could one day assume through resource committees are diverse and include decision-making, monitoring of resources, awareness raising and helping with law enforcement.

Decision-making

The committees could develop site-specific access regulations that more directly address the threats associated with each area. For example, some evidence suggests that the current legal harvest standards set by the government may not be sufficient to prevent the overharvest of wildlife, even if fully enforced. Villagers could therefore vote to adopt stronger management strategies than those required by government regulations should they deem it necessary to protect their food security.

Monitoring of the resource base

Villagers could monitor their natural resources through a village-based monitoring program. This would entail the development and transfer of educationally appropriate monitoring techniques. Analysis of data would initially have to be conducted in collaboration with a conservation NGO, but simply providing villagers with responsibility for data collection would increase their interest and confidence in research results.

Awareness-raising and grassroots campaigns

Committee members could be trained in outreach so that they could communicate conservation goals to the community. Community members are most likely to listen to and learn from trusted community members. Initially the committee could use existing awareness-raising materials. With time, it might produce its own awareness-raising materials based on the local social context.

Law enforcement

Villages currently engage in law enforcement by reporting cases of poaching and hunting camps to the BZP eco-guard unit (Chapter 4). By incorporating local communities into the development of the restricted access system and by strengthening the authority of indigenous communities to manage resources within their hunting zones, law enforcement by eco-guards could eventually take a back seat to village-based management mechanisms.

(4)

Reducing Pressure on Wildlife and Biodiversity

John R. Poulsen[1], Connie J. Clark[1], Paul W. Elkan[2], Sarah Elkan[2], Marcel Ngangoué[3], Pierre Kama[4], Jean-Claude Dengui[4], Jean Ibara[5] and Olivier Mbani[6]

[1]Nicholas School of the Environment, Duke University, Durham, NC
[2]WCS Africa Program, International Programs, Wildlife Conservation Society, Bronx, NY
[3]Société Likouala Timber, Bétou, Republic of Congo
[4]Direction de la Faune et des Aires Protégées, Brazzaville, Republic of Congo
[5]Ministry of Sustainable Development, Forest Economy and the Environment, Brazzaville, Republic of Congo
[6]Wildlife Conservation Society, Brazzaville, Republic of Congo

Landscape conservation is complex because it involves large multi-use areas that incorporate numerous stakeholders with sometimes competing interests. As described in previous chapters, this complexity is exemplified by the situation in northern Congo where timber concessions lie adjacent to a protected area and overlap with the traditional-use zones of indigenous people. Logging operations introduce new threats to pristine forests and their wildlife, and transform the social context by growing the local economy

Tropical Forest Conservation and Industry Partnership: An Experience from the Congo Basin, First Edition.
Edited by Connie J. Clark and John R. Poulsen.
© 2012 Wildlife Conservation Society. Published 2012 by John Wiley & Sons, Ltd.

and attracting immigrants that multiply the pressure on natural resources. Industry is a trigger and accelerator of socio-economic and environmental changes that often break down traditional systems of resource management and leave rural dwellers with little incentive to manage their common property resources sustainably (Brown 2007). The opening of frontier forest also draws the attention of the State. People and resources that had existed outside the reach of government become accessible, resources must be managed, and laws must be enforced.

All these factors coalesce in the development of the bushmeat trade, as the economics of supply and demand transform once benign levels of subsistence hunting into a commercial activity (Box 4.1). The challenge for the BZP was to develop a wildlife management system that: (1) mitigated the negative effects of industrial logging on wildlife; (2) recognized the access rights of local people and strengthened their capacity to manage wildlife populations; (3) encouraged subsistence hunting over commercial hunting as a means of conservation; and (4) protected wildlife populations most vulnerable to local extirpation.

Box 4.1 **Evaluating the 'Bomassa Model' as an effective strategy for PROGEPP (Dr Heather E. Eves)**

Bushmeat in the Congo Basin and duikers as a major source of bushmeat were highlighted as an important issue of inquiry during duiker expert Viv Wilson's *Decade of Duiker Research* (1985–1994) (Grubb, 2002). This was also the time when much of the conservation literature was focused on developing Integrated Conservation and Development Projects (ICDPs) (Western & Wright, 1994), but when challenges to approaches linking conservation behaviors to development incentives were still fairly new (Ferraro & Kramer, 1995).

At the same time, the Wildlife Conservation Society and the government were implementing a core conservation-development strategy as part of the project to manage and conserve the Nouabalé-Ndoki National Park. As part of this strategy it was necessary to gather baseline socio-economic, hunting and duiker ecological data to compare the impacts and outcomes of this strategy. The project needed to test

the comparative impacts of bushmeat hunting in villages where economic alternatives are available, hunting is controlled (no hunting of
endangered species, no hunting with snares, no hunting for commercial
export) and community services are provided (education and healthcare). To assess the impacts, sites under the conservation-development
strategy were compared to other village types, including logging villages
where economic alternatives and community services were provided
without hunting controls and villages where there were no hunting
controls, no economic alternatives and no community services.

The results of this research revealed essential baseline support for the
strategies being employed in Bomassa and Bon Coin (villages adjacent
to the park) at the time (Eves & Ruggiero, 2000). The villages where a
combination of strategies (alternatives, services, hunting enforcement)
were agreed upon by all stakeholders and applied had higher duiker
densities, improved household income and adequate protein supplied
by the available wildlife.

Logging villages (those associated with the Kabo logging concession
that later became the site for PROGEPP) consumed wildlife at a
significantly higher rate and had lower duiker densities close to the
village, despite the offer of alternative sources of income (and protein).
Villages with no income alternatives and no enforcement consumed the
same amount of bushmeat as villages where hunting was controlled, but
had lower densities of duikers close to the village.

The results of this research provided a first account of the potential
positive impacts of an integrated conservation and development strategy
that linked wildlife law enforcement to social benefits (income alternatives, healthcare and education). It also demonstrated that, without
enforcement, the sustainability of hunting was highly unlikely (even
where income alternatives were made available). This research also
revealed the first published data on elephant poaching in the bushmeat
trade, an activity that was widespread among villages where there were
no controls on hunting or development alternatives.

The BZP attempted to meet all of these challenges through a multi-pronged
approach to biodiversity conservation. The wildlife management system was
based on four key approaches:

1. increase in the awareness and involvement of communities in wildlife management;
2. development of economic and protein alternatives to hunting and bush-meat;
3. promotion of selective hunting through law enforcement; and
4. adaptation of management strategies to on-the-ground circumstances.

These can be encapsulated in four activities: community outreach, alternative activities, law enforcement and monitoring/research. This chapter describes these activities, the role of the partners and summarizes results and lessons learned from this four-pronged approach to wildlife management.

Conservation education and awareness raising

Wild game is important to the livelihoods of rural populations in Congo because wild meat or bushmeat is the primary source of protein for most rural people, and hunting can constitute a source of revenue for forest-dependent families. Because their livelihoods are the first to be affected by the overharvest of bushmeat, local people have the most to gain from conservation. They also have the most to lose should conservation efforts fail.

To date, however, many forest-dependent communities in northern Congo have not experienced the effects of overhunting observed in other forested regions. (The area around the town of Pokola is an exception; see Chapter 6.) The idea that wildlife is a limited resource is still not widely accepted. Furthermore, because the concepts of environmentally appropriate wildlife harvest and endangered species protection conflict with the maximization of short-term financial gains from the bushmeat trade, forest-dependent communities perceived conservation and protection as undesirable constraints on their activities.

To assist local people in recognizing the potential negative consequences of unsustainable wildlife harvest and encourage local communities to adopt environmentally appropriate harvest techniques, the BZP initiated a con-servation outreach and awareness-raising program as a priority activity early on. This program helped garner local support for and participation in BZP conservation initiatives by highlighting the long-term benefits of conservation.

The BZP education program promotes conservation and management of natural resources through two main methods: awareness-raising in local communities and environmental education in schools and nature clubs. Activities are broad in scope and employ many outlets including village meetings, theatre groups, radio and television broadcasts, poster campaigns, schoolroom visits and after-school nature clubs. In the early phase of the project, BZP activities were largely geared toward increasing awareness of Congo wildlife laws (Box 4.2) and endangered species in local villages and in CIB towns and camps. (Although wildlife and hunting laws were on the books, the government did not actively inform the public prior to project implementation.) Later, education and awareness-raising activities focused on increasing the understanding of more complex conservation principles such as strategies for sustainable wildlife management to guarantee bushmeat for both present and future generations, and the role of tropical forests for global ecosystem services (water cycle, climate regulation, etc.).

Box 4.2 **Congolese hunting laws**

Congolese Law 48 defines the protected status of game species and sets the rules and regulations for their harvest. For example, endangered species such as elephants and gorillas are *protected* from hunting; species such as forest buffalo with low population numbers are *partially protected* and can only be hunted with a big game hunting license; and non-endangered species, including most species of small antelopes and monkeys, are *unprotected* and can be hunted with a small game hunting permit. The small game permit costs $3.00. Medium and large game permits are more expensive and allow hunting of partially and protected species in certain areas. Hunters are required to purchase a hunting permit at the opening of the hunting season; all firearms used for hunting must be registered with MEF. The hunting laws also delimit the period from November through May as a non-hunting season, during which time only traditional weapons (e.g., hunting nets, spears and crossbows manufactured from natural forest resources) can be used. Finally, the hunting laws define quotas, that is, the number of animals of each species that can be hunted during a single outing and over a season.

Awareness-raising in the community

Village meetings usually begin with the exposition of a poster or video on a given topic followed by a facilitated discussion between the BZP outreach team and community participants. To stimulate interest and reach more people, posters and brochures are generally distributed in village meeting areas (bars, stores, CIB canteens, etc.) following meetings. Posters and brochures are generally printed in both French and local languages. For many campaigns, versions have been printed using only symbols and pictures so they can be easily understood by illiterate community members. Materials designed around the campaign '*Ba Nianma Minene Ya Nord Congo: To Batela Ba Niama Na Bisso*' (The Large Mammals of Northern Congo: Protect our Wildlife) have been particularly successful. Community collaborators and project educators posted images and fun facts about Congo's megafauna at local stores, bars, hospitals, schools, offices and restaurants. A mural based on this campaign was also painted on a wall of the airport so that it was one of the first and last images that travelers saw of northern Congo. Such public campaigns aim to increase the community's sense of pride in their native species and encourage the protection of local biodiversity, while reminding visitors of the local commitment to wildlife conservation.

Themes of community meetings change regularly; examples include the following:

- awareness of protected species (Figure 4.1);
- principles of long-term wildlife management;
- environmentally appropriate hunting techniques;
- hunting permits, seasons and regulations;
- traditional hunting rights of forest people;
- illegal export of bushmeat;
- measures of prevention for Ebola and other emerging human-wildlife diseases;
- development of alternative protein sources;
- role of eco-guards;
- rights of civil society with respect to law enforcement;
- relationship between CIB, BZP, government and community partners;
- rationale and importance of land-use planning/zoning activities;
- tropical forest ecology;
- land-use change and agricultural practices;
- past, present and future rights and practices of semi-nomadic forest peoples.

Figure 4.1   **Two village elders show off gorilla photos they received as part of a Buffer Zone Project poster campaign to raise local awareness of protected species. Photo by Joy Ferrante.**

Radio, television, theatre and game show productions

Weekly radio and TV outlets are used to transfer messages to large groups of people. Because these media are based in CIB logging villages the audience is mostly CIB workers and village residents, and they are effective at reminding workers of the company regulations for hunting and transport. Over the years, however, BZP educators worked to adopt innovative ways of raising awareness of conservation issues among local communities, with a focus on increasing the active participation of adult villagers in these activities.

For example, the BZP organized a traveling theatre group to animate complex conservation issues. One skit demonstrated the importance of law enforcement by depicting the consequences of an unchecked commercial bushmeat market on a small village. Another skit dramatized elephant crop damage in local fields, with the message that the BZP could work with villages to find solutions to human–wildlife conflict.

Game shows   Reward-based game shows have also been popular with teenagers and adults. Villagers compete with one another to answer questions about conservation and endangered species. Questions are based on information provided in village meetings. Winners receive small prizes such as pictures, posters of animals or hats or T-shirts with conservation themes.

Conservation films   Films have been developed in collaboration with International Conservation and Education Fund (INCEF), an NGO that organizes the creation and presentation of conservation and development films. A Congolese film team created the films, which play in local languages and include several different themes such as: ecology of elephants and gorillas; role of eco-guards; human-elephant conflict; sustainable hunting; and Ebola. An average of 250 people participated in village meetings when INCEF films were integrated into the outreach campaign (as compared to an average of 45 people before the introduction of theatre and film). After the success of the films, DVDs were given to the CIB TV station in order to reach a broader audience.

Campaigns targeting CIB employees   In addition to the radio and television broadcasts, the BZP has created conservation-related brochures that are included in the pay stubs of nearly 2000 CIB workers. Meeting with the CIB workers' unions has also proven to be an effective way of building the conservation knowledge of employees. In these meetings, the BZP presented data to union members regarding hunting violations by employees and the abundance of bushmeat in markets. Showing results of monitoring efforts has made employees aware of the environmental consequences of their actions.

Environmental education in schools and nature clubs

Environmental education is important in helping to shape children's values, perspectives and understanding of the environment. In rural communities of Central Africa, most children have had little or no meaningful exposure to environmental education. Because today's children will be responsible for making decisions about future resource use and environmental stewardship, the BZP conservation and outreach program has developed a series of activities in schools and nature clubs dedicated toward fostering a strong conservation ethic among children residing in CIB concessions.

The BZP launched this program in two pilot schools, and quickly expanded it to 27 schools. During its early phases, BZP educators would regularly attend classes to engage students in activities related to forest and wildlife ecology. Later BZP educators began teaching material from the book *The Endangered Species of Congo* written by WCS in collaboration with local schoolteachers to encourage interest in the behavior and ecology of endangered species (Clark & Elkan, 2004). As the school program advanced, an emphasis was placed on engaging schoolteachers in the curriculum. The BZP educator's role shifted to one of teacher-training, organizing and providing teaching materials (e.g., posters, paper and crayons) and encouragement of teachers to continue independently of the presence of BZP educators.

To facilitate the development of this extended education program, the BZP held a series of teacher-training workshops on environmental education and the creation of weekly lesson plans for a range of age groups. At the end of these workshops, *The Endangered Species of Congo* books were distributed to participating schools and teachers taught the material as part of their science curriculum. BZP educators visited the schools frequently to assist the teachers and assess progress in the classroom. BZP educators also created an activity guide with activities (age-appropriate games, puzzles and writing exercises) to stimulate the children's interest in species conservation. Additional teaching aids such as puppets, posters and games were used to engage children. For example, borrowing from the concept of baseball cards, BZP created animal cards that show a picture of a species with interesting facts about its physiology, morphology and ecology. Cards with maps of hunting zones have also been developed so that students are aware of hunting zones and the reason behind the development of zones.

In general, the environmental education program in schools has been successful. During an evaluation of the program by the regional ministry of education, Kabo teachers were commended for their efforts to advance knowledge of biology and conservation and regional authorities requested that the curriculum be expanded nationwide. Nevertheless, environmental education remains impeded by lack of resources; for example, the number of books has not been adequate to meet the demands of all schools and teachers.

Nature clubs

In addition to the program in local schools, BZP hosted a weekly nature club for children ages 5–12 (Figure 4.2) where an average of 45 children participated in a variety of indoor and outdoor games, art projects, puzzles and reading.

Figure 4.2   (Above) Jean-Claude Metsampito surrounded by children at a Nature Club session. (Below) Children participating in the 'hunting' game.

The children are encouraged to explore their natural environment; each session of the nature club has a theme, from learning about local animals and plants to erosion, the water cycle (Figure 4.3), geography and habitats. Specific activities accompany each theme so that children learn through experience: seeing, hearing, smelling and touching. As part of the games and quizzes, children win pictures of protected species and other small prizes for correct responses to questions.

Figure 4.3   The Buffer Zone Project education team instructs Kabo children about the importance of forests to the water cycle during a Nature Club session. Photo by Joy Ferrante.

Role of BZP partners in raising awareness of conservation

In addition to the activities led by WCS, the MEF has conducted yearly village meetings at the opening of the hunting season to remind people of the hunting laws and to inform them of how and where to buy hunting permits. CIB has facilitated BZP awareness-raising activities by allowing free access to all CIB outreach tools (radio and television) and has attached printed educational material developed by the education team (e.g., fun facts about endangered species, list of hunting regulations) to employee pay slips. CIB has also displayed conservation posters in offices, canteens, hospitals and other areas frequented by CIB employees.

Results and lessons learned from awareness-raising

Awareness-raising and environmental education have successfully engaged community members in discussions of conservation. Surveys and quizzes given before and after village meetings and environmental education classes have demonstrated that people came away with an increased knowledge of endangered species, hunting laws and conservation principles – at least in the short term. Perhaps more importantly, attitudes have changed over time. People now talk openly about the importance of the environment and of conserving resources. There is a much greater knowledge of hunting laws and, in particular, the identity and importance of endangered species.

Anecdotally, these observations indicate a change in attitude toward nature, particularly on the part of young people. At one village meeting a prominent community member related how, after participating in environmental education classes in school, his children had posted pictures of endangered animals in their home. The man said that his children insisted he stop buying meat of these species, which led several community members to laugh and exchange similar stories. Interactions such as these provide qualitative evidence that environmental education can cascade to the decision makers (those who buy and prepare meat for dinner).

From a management perspective, however, robust data and quantitative results are most useful for tracking the success of awareness-raising efforts. One would like to know if the money and effort expended for awareness-raising results in a detectable reduction in poaching and a long-term increase in knowledge and change in attitudes toward the environment. The effect of awareness-raising on poaching could be determined by comparing the number of endangered species in bushmeat markets in villages where varying levels or different techniques of awareness-raising had been employed. The long-term effect of awareness-raising could be assessed by surveying a random sample of people on conservation issues over time.

Developing economic and protein alternatives to hunting and bushmeat

The goal of developing economic and protein alternatives to hunting and bushmeat is to guarantee food security while reducing the hunting pressure on wildlife. Currently, nearly 45% of all animal protein in household diets of logging towns is derived from hunting and bushmeat (Poulsen *et al.*, 2009); a level of hunting that may not be sustainable in the long term. In the search for alternative activities to hunting, the BZP has experimented with several types of animal husbandry and micro-projects.

Role of BZP partners in developing economic and protein alternatives

CIB was particularly involved in the development of alternative activities as part of its commitment to make adequate food resources available to its employees. CIB's activities focused on company employees, whereas BZP's activities targeted local, traditional villages (Table 4.1). The government of Congo contributed to the alternative activities program by assigning an agricultural technician to BZP.

Table 4.1 **Results of the alternative activities program in 2004 and 2005. Similar results were produced from 2000 to 2005, but BZP activities (animal husbandry, gardening, provision of fishing materials and chicken farming) were stopped in 2005 due to a lack of financial resources and failure to produce significant quantities of meat and revenue.**

Activity	Jan–Jun 2004	Jul–Dec 2004	Jan–Jun 2005
Animal husbandry (animals given to local people)	51 guinea pigs	11 guinea pigs	12 guinea pigs
Importation of cattle (tons of cattle imported)	18.8 tons	18.9 tons	16.9 tons
Fish farming (kg of tilapia produced)	335 kg	871 kg	286 kg
Gardening (kg of seeds distributed)	6.5 kg seeds	1.3 kg seeds	3.8 kg seeds
Importation of frozen meat (tons of meat imported)	41.1 tons	48.3 tons	23.9 tons
Provision of fishing materials (materials sold to fishermen)	3000 m nets 2000 fishing hooks 18 rolls of line	500 m nets 300 fishing hooks 3 rolls of line	9100 m nets 2400 fishing hooks 41 rolls of line
Chicken farming (materials given to chicken farmers)	885 chickens vaccinated 375 m chicken wire	3769 chickens vaccinated 225 m chicken wire	1377 chickens vaccinated 625 m chicken wire

The BZP furnished material and technical assistance to local people with the following aims.

- Increase chicken and egg production by providing technical advice on raising chickens in coops, by provisioning chicken farmers with fencing materials and by vaccinating chickens against Newcastle disease.
- Increase the production of vegetables by providing local gardeners with vegetable seeds.
- Increase the sources of animal protein through pilot projects with local people by testing and promoting, if successful, the rearing of non-traditional livestock such as snails, rabbits and guinea pigs.
- Increase the production of sheep by giving families in several villages a pair of sheep, with a formal agreement that the first offspring would be given to another family to start their troop and so on.
- Increase fish harvest by providing fishermen with supplies at low cost (prices of materials in Brazzaville or Douala).

Most of these activities have however been unsuccessful in increasing total animal protein for local people. Although people welcomed materials and technical training, they did not easily integrate the new techniques into their daily activities and most household-level projects eventually failed (Elkan *et al.*, 2006). For example, sheep husbandry met with little success because people failed to follow technical advice, resulting in high mortality and low reproductive success (Box 4.3). Local people did not adopt new types of domestic animals such as rabbits, guinea pigs and snails (which have proven successful in West Africa). Either the husbandry techniques required too much maintenance (e.g., keeping animals in cages), or some people did not like the meat. In the case of snails, even though indigenous people occasionally consume snails, other local people were offended by the idea and complained that the project was culturally insensitive.

Box 4.3 **Obstacles to the success of alternative activities**

Environmental constraints: The rainforest is a difficult environment in which to raise domestic animals that have not evolved defenses to tropical pests and diseases (e.g., tsetse flies and trypanosomiasis).

Likewise, agriculture is made difficult by pests and poor soils and often requires clearing the forest using slash-and-burn techniques. Even then, successful crops are susceptible to destruction by forest animals such as elephants and forest pigs (human-animal conflict).

Cultural constraints: Culture contributed to the failure of alternative activities in northern Congo. Domestic animals are usually perceived as savings and insurance rather than sources of protein. They are sold, for example, when a family needs money for medicine or education. Most people do not have a tradition of animal husbandry and agriculture and are used to living off the forest. When resources are abundant, hunting, fishing and gathering can take less time and less physical labor than agriculture. Local people also distrust the unknown. When chickens died after being vaccinated (of a different cause), many people attributed it to the inoculation and refused future vaccinations. Other people declined from the outset, believing outsiders were trying to poison them. Yet other people claimed that the vaccination made the meat taste bad.

Compared to the other activites, selling fishing gear at cost may have increased food availability for local people through greater fish harvest: over 18,600 hooks and 11,600 m of fishing nets were distributed (Elkan *et al.*, 2006). While intensification of fish production is an alternative to hunting, the effects on fish stocks need to be monitored over time to ensure that one unsustainable activity is not being traded for another (see Box 4.4 for such a monitoring program). In terms of increasing food and protein availability, the statistic that matters for alternative activities is not how many seeds, fishing materials or sheep were distributed, but rather the quantity of additional protein and revenues resulting from these activities. The alternative activities program was not very successful from this viewpoint. In terms of building community support for conservation, however, the alternative activities program was crucial. By providing basic material and technical assistance, the BZP demonstrated concern for the livelihoods and wellbeing of local people. These activities helped to communicate the message that conservation is not just about saving gorillas, for example; it is also about managing resources for the long-term use of local communities.

The production of protein has also been attempted in logging towns through activities led by CIB with technical assistance from the BZP. Early in the project,

CIB and BZP established two large fish farms, four chicken farms, two butcher shops, one slaughterhouse and five cold rooms to store imported meat (pork, chicken and seafood). CIB provided the supplies and manpower to build the different structures (e.g., bulldozers were used to dig the fishponds). Despite targeting motivated individuals with good work records (most were CIB employees), the fish farms failed to produce a substantial harvest. Dependence on imported chicken feed was logistically difficult and, without a local alternative, the chicken farms were not cost-effective and were closed.

To increase the amount of available animal protein in logging towns, CIB has imported frozen meat and assisted local tradesmen in the importation of cattle and frozen meat every few months. This has been the most productive activity in terms of increasing the meat supply. Between 2002 and 2006, an average of 79,200 kg (standard deviation or SD = 15,565 kg) of frozen meat and 29,772 kg (SD = 10,440 kg) of beef was imported into the four forestry towns (Pokola, Kabo, Ndoki 1 and Ndoki 2). Even so, the frequency of domestic meat in meals peaked at about 15% in the four villages for any particular month (Poulsen *et al.*, 2009). Annually, 5–7% of all meals has contained domestic meat in the logging towns, primarily as a result of this effort. Two different opinion polls conducted by BZP found that local people prefer to eat beef, chicken and fish to some types of bushmeat (BZP, unpublished data). The results of these polls imply that residents would likely welcome increased production or importation of domestic meat, fish and poultry.

Contribution of alternative activities to protein supply and household budgets

Poulsen *et al.* (2007) used back-of-the-envelope calculations to estimate the proportion of concession residents whose daily protein requirements would be met by the meat importation program. They estimated that the importation of nearly 100,000 kg of protein per year provided 11.4% of the needs of company employees and their families and 4.8% of the needs of the entire population of the concessions. The amount of imported meat would have to be increased threefold to satisfy the needs of the residents of just the Kabo and Pokola concessions. Vautravers (2008) estimated that the importation of meat into the concessions was over 200,000 kg in 2007 (95,228 kg of beef and 110,088 kg of frozen meat) and could provide 24% of employees with protein.

The alternative activities program and CIB's efforts to provide domestic protein broke dramatically from the typical practice of feeding logging employees through hunting (Robinson *et al.*, 1999). Even so, a great deal more work needs to be done to ensure food security for all concessions residents, not just company employees. The standard for industry should be to provide 100% of protein for its employees. If 24% of the protein needs of employees comes from domestic meat, another 76% must be met through bushmeat and fishing. By relying on wild meat company employees put stress on wildlife resources, potentially compromising the long-term food security of local people native to the area.

There are few, if any, successful alternative activities projects in Central Africa (Box 4.3). Indeed, BZP did not reach its goal of providing a consistent alternative to hunting by producing substantial revenue or protein for local people (Poulsen *et al.*, 2007). Most of the work on alternative activities was conducted between 1999 and 2004, after which the program was scaled back due to lack of funding. The rainforest is a difficult environment in which to raise domestic animals that have not evolved defenses to tropical diseases. For these reasons, animal husbandry is not considered a viable option for food security (Vautravers, 2007). Likewise, agriculture is made difficult by pests and poor soils, and often requires clearing of the forest through slash-and-burn agriculture. The people of northern Congo do not have a tradition of animal husbandry and agriculture. When resources are abundant, hunting, fishing and gathering can take less time and less physical labor than agriculture. Overcoming these obstacles will require expertise, resources and time (Box 4.3).

Potential solutions for feeding concession residents

Based on a very limited study, Vautravers (2008) estimated that the Sangha River and Terre de Kabounga fisheries could provision all CIB employees and their families with adequate fish. The distribution of fish to population centers such as Pokola currently fails due to poor fish preservation techniques and lack of an easy means of transport. Among several possible solutions, Vautravers (2008) proposed that company boats be dispatched weekly to fishing camps, that fisherman be given free transport on company vehicles, that local merchants be paid to serve as the link between camps and markets and that small rivers be opened up for easier travel by canoe to markets. Preservation of fish could be improved through the construction of refrigerated chambers for fresh and smoked fish.

Although these recommendations seem practical for provisioning logging towns with freshwater fish, they ignore ecological and socio-economic concerns. As Vautravers (2008) recognized, the size of fish populations in the rivers in and around the concessions is unknown. Whether or not harvest levels are sustainable is also unknown (see Box 4.4). An outstanding issue is how the distribution of the fishing catch to Pokola would affect local villages. Will unemployed local people be left without adequate protein? Would this be another case where company employees benefit from local resources for which local people should have priority?

Box 4.4 **Sustainable fisheries as an alternative to bushmeat hunting (Hannah L. Thomas, Wildlife Conservation Society)**

To reduce hunting pressure on Congo's diverse wildlife, alternative sources of protein need to be found. The Sangha River is one of the major watercourses in the Congo Basin; it provides an important supply of fish to local populations, particularly in the dry season (January to March) when low water levels favor fishing. Livelihood data collected by BZP demonstrate that bushmeat consumption is negatively correlated with fish consumption. The Sangha River fisheries could potentially provide a viable alternative to unsustainable bushmeat offtake. Until recently, however, neither the diversity of fish species in the Sangha River nor the level of fishing pressure upon the fisheries was known.

The Sangha Tri-national Landscape (STL) Fisheries Project is a transboundary venture initiated by WCS in the Republic of Congo, in collaboration with WWF in Central African Republic and Cameroon. The immediate goals of the project are threefold: (1) to improve knowledge of fish diversity; (2) to assess the levels of fishing offtake; and (3) to support local communities in sustainable fisheries management. To assess fish diversity, 400 km of river was stratified into four zones (Figure 4.4) according to the economic influence of nearby commercial centers. Researchers from all three countries used a common stock assessment methodology, placing gillnets and capturing and collecting fish at over 40 sites on a monthly basis for 10 months.

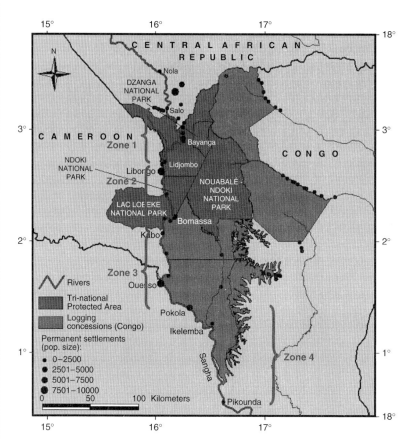

Figure 4.4   **Map of the Sangha Tri-national Landscape and the four fishing project zones.**

Ichthyologists from the Institute of Rural Development at Brazzaville University identified all specimens.

Preliminary results from the study found 113 different fish species in the study region; and this number is likely to increase as the classification of unidentified specimens is completed. There were marked differences between species richness among geographic study strata. Stretches of river flanked on either side by protected areas harbored higher species richness (54 and 77 species for Zones 1 and 2, respectively) than areas surrounded by logging concessions and next to commercial centers such as Kabo, Pokola and Ouesso (18 and 28 species for Zones 3 and 4, respectively). A similar pattern occurred for commercially valuable fish species. The number of commercial fish species was highest in Zone 2 (16 species) compared to 6, 5 and 5 species in Zones 1, 3 and 4, respectively. Because protected areas in the STL are free from human habitation, these results suggest that overfishing may be seriously threatening fish biodiversity in certain stretches of the Sangha River.

The second phase of the project, assessing the levels of offtake in each zone, is now underway. Average catch-per-unit-effort numbers will be used to compare offtake among zones. The final phase of the project aims to develop and strengthen local fishing committees. By building management capacity at a local level, committees will be able to implement resource-use decisions at ground level. The CIB logging company has pledged support for the project through the facilitation of transport and refrigeration to open up accessibility to fish stocks throughout the year. Although this project is greatly advancing the potential to manage fish resources for sustainability in the Sangha River, the scale of the bushmeat crisis in northern Congo means that management of fisheries by itself is unlikely to provide a complete solution. Finding additional livelihood and protein alternatives is still an urgent priority if we are to maintain wildlife populations in the Congo Basin.

CIB has made laudable efforts to provide food for its employees but attempts at raising crops and livestock have not been productive, at least not at the industrial scale necessary to feed 12,000 people. Although the company was only given a lease for the harvest of timber, it continues to profit from both the fish and wildlife resources of northern Congo (to the detriment of local people).

Promoting selective hunting through law enforcement

While the access regulations that determine who can hunt and where (Chapter 3) were largely created by the BZP partnership and local communities, wildlife laws regulate legal wildlife harvest[1]. Like many Central African countries, Congo has well-defined hunting laws that determine what (species and number), how (techniques) and when (seasons) people can hunt (Box 4.2). Insufficient financial and human resources limit the ability of government to enforce its laws in the field, however. In contrast to most protected areas where eco-guards are employed to protect biodiversity, forests outside of parks and reserves usually lack any enforcement. The BZP was the test case for using eco-guards to enforce hunting laws and control the transport of bushmeat in timber concessions. As a result, the 2002 Congolese forestry law required logging companies to pay for a law enforcement unit (Unité de Surveillance de Lutte Anti-Braconnage, USLAB) within their concessions. The goal was not to completely restrict public access to logging concessions (as some authors argue, e.g., Brown, 2007), but rather to use the resources of logging companies to enforce laws that are meant to be implemented nationwide. Many companies have been reluctant to engage fully in this process, and thus wildlife laws are not enforced in concessions nationwide.

As part of its role in the BZP, CIB integrated Congolese hunting laws into its company rules (Elkan & Elkan, 2005). In so doing, it made a commitment to educate its employees about the rules and to enforce them through fines, loss of work hours and other penalties. The company rules seek to maintain biological diversity in the concessions, protect endangered species, minimize the commercial harvest of wild animals and reduce the indirect impacts of logging on the Nouabalé-Ndoki National Park (Congolaise Industrielle des Bois, 2006). CIB has incorporated standards that exceeded the national law (Box 4.5). The restriction on exporting bushmeat from one site to another is not part of Congolese law, although the Minister of MEF articulated it as a statement of policy.

Box 4.5 Congolaise Industrielle des Bois company rules on hunting

1. The hunting of protected species and use of snares made of metal or nylon are prohibited.
2. Workers must obtain a hunting permit and license to carry a firearm.

---

[1] To date, legal levels of wildlife harvest enforced at BZP are dictated exclusively by national wildlife codes. However, the BZP project aims to move toward the development of community-based management plans for wildlife that would regulate the level of harvest at smaller spatial scales.

3. The transport of hunters, firearms and bushmeat in company vehicles is prohibited.
4. Drivers are responsible for the people and materials transported in their vehicles; drivers can therefore be penalized if they carry bushmeat or hunters.
5. Drivers must stop at control posts and permit eco-guards to search vehicles.
6. Driving at night without written authorization is prohibited.
7. Land-use zoning must be respected; protected and conservation zones are off limits to hunting.
8. The export of bushmeat outside the zone where it was captured is prohibited (i.e., only local consumption of wild meat is allowed).

In most cases, employees who break company rules for the first time receive a written warning. Second and third violations result in unpaid suspension from work for 1–8 days and the loss of year-end bonuses. The fourth violation results in the loss of employment.

Poaching of a protected animal species, considered the most serious violation, results in immediate dismissal from the company.

Paramilitary guards enforce the national hunting laws and company rules. These eco-guard teams police the concessions through targeted forays into the forest, searching for poachers and snares in areas thought to be threatened by illegal hunting, and through inspections of vehicles at roadside posts along the logging road network.

In the event that eco-guards apprehend a poacher or driver (Figure 4.5), they confiscate any illegal animal products or unlicensed weapons and write a 'note of infraction' describing the violation. Along with the confiscated materials, the note is transferred to and reviewed by the MEF agent in charge of the eco-guard unit. If warranted, a ticket is issued. Every month, these tickets and confiscated materials are sent to the MEF brigade chief. Gun owners can reclaim their guns from the regional office by paying a fine (unless the weapon is an illegal one such as a PMAK automatic weapon). MEF agents sell confiscated bushmeat in local villages, with proceeds being divided between the functioning of the eco-guard unit and the regional MEF office. If an

Figure 4.5 **An eco-guard prepares to search a logging truck for bushmeat and weapons. Eco-guards man roadside posts and search all vehicles that pass through the CIB concessions. Photo by David Wilkie.**

endangered species has been killed, the poacher is delivered to the regional jail to be tried and sentenced[2].

BZP eco-guards have not enforced hunting laws to their full extent because of insufficient manpower and logistical resources and the impracticality of some laws. For example, stopping all hunting with modern weapons during

---

[2] In reality, very few poachers ever go to trial or face sentencing. After seven years, only one repeat elephant poacher was tried and sentenced to six months in the Brazzaville penitentiary. All other poachers were freed from the regional jail without any formal explanation.

the non-hunting season (November through April) would leave concession residents (including indigenous peoples) with insufficient animal protein. For similar reasons, daily and seasonal quotas have also not been fully enforced (although a hunter found to possess more than four animals is likely to be apprehended on the basis that Congolese wildlife laws forbids commercial hunting). Finally, the hunting of partially protected species, for which a special hunting permit is required, has often been overlooked.

As demonstrated above, the law enforcement unit is flexible when it comes to the enforcement of subsistence hunting. Several aspects of the law have been strictly enforced including the prohibitions against hunting with snares, the hunting of protected species, the transport of bushmeat on logging vehicles and compliance with the hunting zone system. Also, all drivers or company employees that break the rules are ticketed without exception. Depending on the severity of the infraction, CIB employees can be given a verbal or written warning, put on leave without pay or fired.

Roles of BZP partners in law enforcement

Law enforcement is usually conducted by national governments alone. However, in several Central African countries international NGOs provide technical assistance and logistical support to law enforcement efforts. In Congo, WCS partnered with MEF to manage four protected areas in addition to the BZP, and one aspect of this management included law enforcement.

Three circumstances dictated the level of WCS's involvement in law enforcement: the administrative capacity of MEF, the ability of the MEF coordinator to lead, and the severity of poaching. At first rules and administrative procedures for law enforcement were poorly defined, requiring greater assistance from WCS. For example, until 2008 eco-guards were hired as WCS employees rather than as government employees. This changed when the government established the necessary administrative structures to hire them. MEF leaders sometimes lacked leadership skills, and the WCS PTA had to help build a team of eco-guards motivated to do a rigorous job. From 1999–2001, the high level of poaching (particularly of elephants and apes) required WCS to engage actively with the MEF to end the crisis.

In the early stages of the project, enforcement of hunting laws was at times met with serious resistance from some local people as well as professional poachers. The modification of the CIB company rules required over 25 hours of negotiation with the employee union (Elkan & Elkan 2005; Elkan *et al.*, 2006). On a few occasions death threats were made against eco-guards, the MEF coordinator and patrol leader and the WCS project staff.

When project personnel were threatened, efforts were made to make high-level MEF leaders (e.g., Director of Wildlife and General Director of MEF) and CIB directors aware of the risks and to mobilize their support. In some cases, the CIB leadership summoned influential union members to ask for assistance in alleviating tension with company employees. In addition, to demonstrate government support for law enforcement, regional-level authorities from the court system occasionally visited the BZP and warned individuals not to interfere with wildlife law enforcement efforts. These measures helped alleviate tension and, over time, CIB employees became accustomed to law enforcement measures.

At the site level, the MEF and WCS worked together to develop anti-poaching strategies and to guide their implementation in the field. Typically, a long-term law enforcement strategy was designed with the aid of socio-economic, ecological and law enforcement data. The long-term strategy was modified as necessary in the implementation of six-month and monthly work plans (see *How does it run?* below for more details). In general, collaboration worked well between the MEF and WCS partners in the face of a difficult and complex law enforcement situation.

With passing time and awareness-raising campaigns, law enforcement has become an accepted part of life in the CIB concessions. In addition, MEF and its agents have grown more experienced in personnel management and law enforcement as the wildlife management system has become more refined. The role of WCS has evolved accordingly to that of a monitoring and advisory role with MEF as the leading partner in management and implementation of wildlife law enforcement.

The logging company's principal contribution to law enforcement has been through financial and logistical support. CIB contributed $10,000 per month for eco-guard salaries, furnished the project with two trucks and drivers and housing for most project employees and built and maintained roadside posts and barriers. CIB managers and employees have assisted law enforcement by disclosing information about poaching to the BZP management team. CIB has also been responsible for enforcing its company rules, which entailed keeping track of employee infractions and sanctioning employees who broke the rules. On multiple occasions CIB managers turned in company employees when they were found illegally hunting or transporting bushmeat. One observant CIB forest manager became curious when he spotted blood leaking from a bulldozer. With some probing, he discovered the carcass of a forest pig hidden in the exhaust manifold, which he confiscated from the driver and turned over to eco-guards. The driver was ticketed and suspended from work for several days.

Eco-guard staffing levels

In the CIB concessions, between 25 and 40 eco-guards enforce the hunting laws and company rules. Determining adequate levels of enforcement has proven to be a challenge. Although it is tempting to specify a ratio of personnel per unit of land, such prescriptions are useless unless they are based on the hunting pressure and strategic landmarks in an area. For example, the NNNP (426,000 ha and 0 inhabitants) – where no evidence of poaching was observed between 2005 and 2007 – does not require the same number of eco-guards as the Kabo concession (296,000 ha and ~4200 inhabitants), where 10,489 snares were confiscated in 2005. When estimating the number of eco-guards to employ, considerations must include indicators of hunting pressure such as the number and length of roads, number of villages and towns and human population. In practice, the number of BZP eco-guards has fluctuated over time, primarily for financial reasons. The project began with 10 eco-guards in 1999 and reached a peak of 45 in 2004. In 2007, 25 eco-guards were determined insufficient to cover the Kabo, Pokola and Loundoungou concessions, and managers closed some roadside posts in order to police the forest areas with the highest levels of poaching.

A starting point for assessing the number of eco-guards is to first identify the number of fixed posts needed to control traffic and villages. Roadside posts make the transportation of hunters and bushmeat more difficult, but hunters learn to bypass them. If possible, fixed posts should be positioned near bridges or inaccessible habitats (e.g., swamps) that are difficult to circumnavigate. Second, the number of ecologically important sites vulnerable to poaching must be assessed. These could include habitats of endangered species, forest clearings that animals frequent for water, food or mineral resources or corridors that provide safe movement of animals from one protected area to another. If the goal is to conserve wildlife populations across the entire area or concession, the strategy might be to send patrols where poaching is highest (probably near towns and villages). Third, the frequency at which sites need to be visited to inhibit poaching must be estimated to determine the number of forest patrols necessary. A final consideration is to determine the optimal size of eco-guard teams. In the BZP, fixed posts are manned by three eco-guards at a time whereas forest patrols consist of five eco-guards. Once the number of eco-guards has been calculated on strategic considerations, the number should be increased by 25% to account for illness, leave and other absences of personnel.

Costs of law enforcement

Law enforcement is an expensive endeavor. In addition to salaries, bonuses and health benefits, eco-guards must be equipped with a uniform, field and camping equipment and provisions. Because of the rough field conditions in tropical forests, equipment and uniforms need to be replaced annually. Transporting eco-guards long distances requires vehicles and boats which in turn incur fuel, maintenance and repair costs. On average, it costs about $5000 per year to pay, provision and equip an eco-guard for the field in the Congo; this does not include vehicle transport, fuel, administrative or training costs and infrastructure construction and maintenance costs.

Paying the costs of law enforcement is usually incumbent upon the government or company. NGOs and international donors are reluctant to pay eco-guard salaries and expenses, which can be perceived as funding national militias. The government of Congo requires logging companies to finance a unit of 10 eco-guards per concession. Unfortunately, this arbitrary number does not account for the actual pressure on wildlife or the overall size of the concession (it takes more eco-guards to cover 500,000 ha than 500 ha). CIB has consistently paid $10,000 per month toward salaries of the BZP eco-guard unit. At $5000 per eco-guard per year, the CIB contribution can maximally support 25 eco-guards despite being required to employ 30 eco-guards. The company's contribution has not grown in step with population growth (~60% from 2000 to 2007; Chapter 6), expansion of the road network or with the addition of the Loundoungou concession (552,500 ha; 16 km from the NNNP).

Recruiting eco-guards

BZP has hired eco-guards primarily from the traditional villages in the timber concessions. Many known poachers were deliberately hired to give them an alternative livelihood[3]. Hiring locally has had several benefits. Salaries are one way that local people benefit from the conservation of biodiversity other than through its consumption. Even though relatively few people are hired, their salaries trickle back to the village to support families, providing an incentive for village members to abide by hunting laws. Locals tend to be more knowledgeable of forests and hunting practices, making them effective at policing hunting. The disadvantage of hiring locally is that eco-guards might be tempted to turn a blind eye if they confront a relative or friend breaking the law. Social pressure to 'look the other way' can be intense, and more than one eco-guard has resigned because of fears of witchcraft or retribution for doing their jobs.

New recruits have been trained and evaluated during a 3-month basic training program conducted by MEF. Military officials from the regional army base were employed to impart the necessary discipline, paramilitary techniques and field methods. Local and national experts, including BZP employees and MEF agents, have taught courses. The training curriculum includes courses in hunting laws and regulations, orienteering, military procedures and techniques, ecology and human rights. Teamwork and physical fitness were built through physical exercises, tests and field practices. Only recruits that performed at a high level were hired as eco-guards.

How does it run?

Once the eco-guard unit is recruited and trained, the difficult process of enforcing hunting laws begins[4]. As mentioned above, eco-guards primarily perform three functions: roadside posts, forest patrols and mobile vehicle patrols along logging roads. In the CIB concessions, six roadside posts are nearly always occupied by two–three eco-guards with the objective of stopping and searching vehicles to block the transportation of hunters, weapons and

---

[3] The principle of hiring poachers was not always successful as some poachers hired as eco-guards continued to poach either on-the-job, in which case they were appropriately sanctioned, or after they left the eco-guard unit.

[4] BZP eco-guards are armed with automatic weapons. It is rare that eco-guards fire their weapons in the line of duty, but elephant poachers have shot at eco-guards on more than one occasion and the ability of eco-guards to defend themselves is critical to the success of their mission and to protect their lives. The MEF coordinator and brigade chief assign weapons and a limited amount of ammunition to eco-guard leaders every week. A register of the weapons and ammunition is kept to account for every gun and every bullet.

bushmeat. A running log is kept of all vehicles that pass the post, including the time of day and the number of passengers. The posts are positioned at the junctions of the main logging roads where they intercept the most traffic, and outside of logging towns. When vehicle traffic is low, eco-guards make excursions into the forest to look for snare traps, signs of hunting (i.e., fires or gun shells) and poachers.

Forest patrols involve week-long forays of five–six eco-guards to stop poachers, collect wire snares and gather information on the level of human pressure in the targeted area. Two–three patrols were conducted per week, each targeting a different area. The MEF coordinator and MEF brigade chief decided the patrol routes based on several considerations: (1) where logging was active; (2) where endangered species were most abundant; (3) where hunting pressure was routinely high; and (4) new information and intelligence on the location of hunting camps or active poaching. Patrol routes are varied so that the entire landscape is covered every few months. It is not uncommon, however, to repeatedly send patrols into an area where there is heavy and active poaching pressure.

Mobile vehicle patrols along logging roads cover a large area very quickly with a focus on the road system. A team of five–six eco-guards is deployed in a vehicle with the purpose of stopping poachers along the roads and installing temporary road blocks to search logging vehicles at unexpected locations (i.e., not at the permanent and well-known roadside posts). Mobile vehicle patrols are effective because they take poachers and drivers by surprise. One of the strategies employed by these patrols is to sound the horn as the vehicle drives along the logging roads. Because the patrol vehicle is the same type of vehicle as many of the CIB transport vehicles, poachers can mistake them for a logging vehicle that has come to pick them up. The poacher is apprehended as he runs out of the forest, sometimes laden with a sack of bushmeat and a rifle.

In addition to these routine law enforcement activities, rapid deployment missions occur in response to specific real-time information on poaching (e.g., location of a poaching camp, discovery of a dead elephant, etc.). In most situations, an eco-guard team is deployed quickly, employing a well-conceived strategy to locate and arrest poachers. In 2006 eco-guards received word that a load of ivory was to be shipped down the Sangha River to Ouesso, and a team was rapidly deployed under the cover of nightfall. Early the following morning, eco-guards ambushed the transport boat, seizing several sets of ivory and arresting two poachers armed with automatic weapons.

Rapid deployment missions are typically short, punctual, risky and, when well planned and executed, highly effective.

Chain-of-command and rotations in teams and assignments

Like a military unit, an eco-guard unit is structured hierarchically to establish a clear chain-of-command so that the line of responsibility is clear and there is no confusion in decision-making. At the BZP the MEF coordinator is the head of the unit and responsible for law enforcement at the administrative level. The MEF brigade chief is responsible for activities in the field. MEF patrol leaders lead eco-guards teams in the field. The number of MEF patrol leaders has typically been low (only three patrol leaders have been assigned to the BZP at one time), so most fixed posts and forest patrols are headed by experienced eco-guards who have been promoted to the rank of eco-guard leader; they must take an oath to the government to be authorized to write notes of infraction against wrongdoers.

Eco-guard teams and post assignments are changed each week. This changeover occurs on a single day of the week so that roadside posts are unmanned for no more than a few hours. During this rotation, eco-guard teams receive new assignments, turn in reports and confiscated materials, replenish provisions and attend the weekly law enforcement meeting. Modifying team membership and post assignments weekly has the advantages of: (1) encouraging team-building and improving team moral; (2) helping to avoid the development of cliques based on ethnicity or friendship that could cause divisions within the greater unit; (3) allowing for an equitable distribution of choice assignments among the eco-guards[5]; and (4) preventing the formation of relationships (positive or negative) between eco-guards and villagers or truck drivers. Familiarity between eco-guards and drivers or hunters can develop into relationships of collusion and result in corruption.

Incentive system

To reward performance, team members are paid a bonus on top of their monthly salary for confiscated items. For example, the team splits 100 CFA francs ($0.20) for every wire snare that is collected during a forest patrol.

---

[5] Compared to work at roadside posts, forest patrols are more onerous and involve camping and working in very rough conditions. Even so, eco-guards prefer forest patrols because they earn a bonus for working in the forest and patrols usually result in higher rates of confiscations.

The bonus can be a substantial incentive as a single team can sometimes collect tens or hundreds of snares during a single week. Higher bonuses are paid for items that are more difficult to confiscate: a pair of ivory tusks earns 70,000 CFA francs ($140), whereas a shotgun earns 10,000 CFA francs ($20). In designing an incentive scheme, it is important that the bonuses be high enough to motivate the eco-guards but not so high that the eco-guards commit fraud for the bonuses.

Information-gathering network

One of the most important aspects to successful law enforcement is the collection of real-time information on hunting. To stop poaching, eco-guards need to know who is hunting and where and when. The best way to gather such information is through relationships with local communities. On several occasions, villagers have divulged information of the whereabouts of poachers, particularly when outsiders trespass on village hunting territories. Company personnel who spend a lot of time in the forest and on logging roads are also a good source of information. Information can also be actively gathered. One strategy is to set up a network of informants that report to project managers. The drawback is that informants usually expect to be compensated and may also risk their own safety. Protecting the identity of informants is extremely important. Through their relationships with local people and communities, BZP researchers are also often privy to information that eco-guards or MEF officials could not otherwise access. On many occasions, villagers have provided information to researchers that led to the apprehension of poachers.

Challenges to effective law enforcement

Law enforcement is a complicated endeavor and strategies for curbing poaching must be constantly updated as threats to wildlife change. Ironically, as the expansion of roads and construction of cell towers create faster transportation and better communication for law enforcement efforts, they also facilitate the transport of illegal products and the ability to circumvent arrest. However, the greatest challenge to law enforcement is the failure of the justice system to prosecute poachers (Boxes 4.6 and 4.7).

Box 4.6 **Elephant poaching: A challenge to law enforcement**

Challenges to law enforcement are exemplified by the effort to stop elephant poaching, the greatest single problem that faces the BZP eco-guard unit. Elephant poachers are generally professionals that hunt the animals solely for their ivory. With automatic weapons, they kill as many elephants as possible, leave the carcasses to rot and transport the ivory to Ouesso or Brazzaville where it is sold on the black market. Local Mbendzélé are often recruited for elephant poaching because of their knowledge of the forest and skill in hunting. It was rumored that a high-ranking military official in Ouesso operated the elephant poaching ring: these rumors were substantiated by the easy access that elephant poachers had to automatic weapons and munitions. In any single year, BZP eco-guards and researchers found 20–30 poached elephant carcasses: many more were likely never found. Since the inception of BZP in 1999, eco-guards arrested several elephant poachers (some of them several times). The chance to earn more than $200 for a single tusk makes the risk of being trampled by an elephant or spending several days or weeks in jail worthwhile, however. Even if a poacher was arrested, the failure of the Congolese justice system to prosecute poachers means that as soon as the poacher is released from the local prison he is back in the forest shooting elephants. As one eco-guard asked, "After risking one's life to apprehend the same poachers over and over, what is the point?"

Box 4.7 **Problems in the enforcement of laws in the CIB concessions (Jean-Claude Dengui, MEF BZP coordinator)**

The legal prescriptions that guarantee the management of the environment, fisheries and other natural resources (fauna and flora) are defined in Law 48 (April 21, 1983). The law is only weakly enforced, however. It has not achieved its desired results because it is not adapted to the context of local communities, who do not see any benefit in following them. Furthermore, rural communities are not involved in decision-making for resource management. Similarly, the application of wildlife laws in the CIB forest concessions will not be effective without alternatives to commercial hunting as income generation for the rural poor. The illegal trade of trophies (ivory, skins of bongo, hippos and certain monkeys), the failure to punish poachers, the lack of jails in the large cities of the Sangha and Likouala regions, the lack of accountability of political and military officials implicated in the organization of poaching networks also weaken the national law. Finally, the lack of ethics of law enforcement agents, partially a result of their inadequate salaries, also limits law enforcement in the CIB concessions and nationally.

Avoiding corruption

Corruption in an eco-guard unit can spoil the entire operation. Once a single member of the unit gets away with taking bribes, other members may follow suit. Corruption in the BZP eco-guard unit has manifested itself in several ways: (1) eco-guards have been bribed by hunters and traders not to confiscate weapons and bushmeat; (2) MEF agents and eco-guards have sold confiscated bushmeat and kept the profits; and (3) at least two MEF agents have hunted while on duty using government-issued weapons. The only solution to corruption is tight and tireless management. It must be clear that corruption and poor work will be punished with strict penalties and loss of employment – with no exceptions. Because law enforcement takes place in the forest out of view of project managers, tight management is difficult to accomplish.

Several ways to avoid corruption include management procedures such as:

- rotation of eco-guards among different road posts or forest stations;
- rotation of eco-guards among teams;
- random field inspections of teams;
- bonuses for confiscated items;

- imposition of clearly defined penalties for corruption; and
- intelligence from independent sources on poaching and eco-guard activities in the field.

Running a tight ship is difficult, particularly for MEF officials that feel social pressure from local people or who have limited management experience. On several occasions, the WCS technical advisor has argued that eco-guards involved in corruption should be fired while the MEF coordinator has argued for leniency. Cognizant of the limited opportunities for employment in Congo, MEF officials are often reluctant to fire employees even in the face of concrete evidence of inappropriate behavior. Unfortunately, the failure to impose penalties can lead to greater abuses.

Abuse of power

One abuse of particular concern is the inappropriate exercise of power over minority populations. In the past, BZP eco-guards have been accused of threatening and beating up Mbendzélé. The Mbendzélé are a minority ethnic group, a clan of the large Aka peoples sometimes referred to as Pygmy, that traditionally lived as hunter-gatherers. Their forest-based livelihoods prevent them from obtaining much formal education and they tend to be poorer and less well integrated into village life than Bantus. Because eco-guards carry guns, wear uniforms and represent authority, their very presence can intimidate Mbendzélé. Accusations of physical abuse are however difficult to substantiate due to both a lack of physical evidence and because poachers may accuse eco-guards of actions they did not commit; similarly, eco-guards may cover up abusive actions they did commit.

All cases of reported abuse of power or corruption by eco-guards are investigated carefully and systematically, maintaining the premise that the accused are innocent until proven guilty. In several instances, poachers have fabricated corruption charges against eco-guards in order to undermine law enforcement efforts. In other cases, the eco-guards have behaved unprofessionally. It is of the utmost importance that an objective and systematic approach be taken to address allegations and monitor staff performance.

Inappropriate actions by eco-guards weaken the unit and make it less effective. If eco-guards are perceived to be unlawful, locals may not cooperate or may actively resist law enforcement efforts. Allegations of wrongdoing can also threaten project funding. International donors may pull support if they think their funds are supporting illegal or abusive activities. Bad publicity for the logging company could lead to lower sales, difficulty with shareholders and even endanger its certifications.

Weak legal and judiciary system

Besides corruption or inappropriate actions by eco-guards, law enforcement has sometimes been crippled by the failure of the Congolese justice system to prosecute poachers. Even though BZP transferred dozens of poachers every year to the regional jail in Ouesso, not a single poacher was tried or convicted until 2007. Prisoners are not fed or given health care in regional jails. BZP has had to pay from its limited conservation funds to keep poachers in jail while they awaited trial, only to see the prisoners freed without explanation. In 2007 an elephant poacher was tried and sentenced to six months in the Brazzaville penitentiary. The poacher, like many others, had been arrested several times by BZP eco-guards. To our knowledge, this was the first time in the history of Congo that a poacher was sentenced for breaking wildlife laws. Although project management was hopeful that the prosecution of the poacher signified a new willingness by the government to uphold its wildlife laws, no poachers have since been prosecuted.

Law enforcement monitoring: results and conclusions

Law enforcement monitoring (LEM) is critical to evaluating the success or failure of the eco-guard unit and law enforcement strategies. LEM should consist of at least three components: monitoring the effort expended in law enforcement; monitoring the results of law enforcement; and examining both efforts and results of law enforcement spatially through GIS.

Monitoring efforts of law enforcement

Law enforcement effort is monitored over time by recording: (1) the number of active eco-guards; (2) the number of days or hours spent in the field per eco-guard; and (3) the number of days of manned posts or forest patrols (Table 4.2). These data are useful for assessing the overall investment in law enforcement from period to period. They can also serve to compare the relative investment in roadside posts versus forest patrols. Forest patrols are

Table 4.2 **Law enforcement effort in terms of the number of days over which each of the types of patrol were conducted (S1: January–June; S2: July–December) from 2003 to 2006.**

	2003 S1	2003 S2	2004 S1	2004 S2	2005 S1	2005 S2	2006 S1	2006 S2
Roadside posts	966	871	814	1069	1036	1028	1050	987
Forest patrols	127	375	342	528	427	372	198	256
Rapid deployments	7	0	0	2	21	10	20	12
Bi/Tri-national patrols	22	17	0	11	6	0	0	10
Total	1122	1263	1156	1610	1490	1410	1268	1265

also monitored spatially by plotting GPS tracks of the area patrolled: maps of these tracks can be used to ensure that the entire landscape is surveyed (Figure 4.6).

Law enforcement results are monitored by recording the number of tickets written and the number of carcasses, snares or weapons collected or confiscated (Table 4.3). Relating law enforcement efforts to results provides a measure of the return per effort (e.g., number of confiscations per eco-guard or per forest patrol). Return per effort can be used to evaluate hunting pressure and effectiveness of eco-guards. When one site has a higher return than another, this suggests that it is suffering greater hunting pressure. Alternatively, the effectiveness of law enforcement can be evaluated by comparing return per effort over time (assuming eco-guards are consistently effective).

In the CIB logging concessions, the eco-guard unit had a higher return per unit effort in the early years (1999–2002) of the project than during 2004–2006. During the period 1999–2002, 13 high-caliber rifles, 7 automatic weapons, 59 tusks, 650 six shotguns and 25,800 wire snares were confiscated in 3139 patrol unit days (Elkan *et al.*, 2006). For every 100 days of patrols between 1999 and 2002, the eco-guard unit confiscated 822 snares, 21 shotguns, 2 tusks, 0.22 automatic weapons and 0.4 high-caliber rifles. In comparison, for every 100 days of patrols during 2004–2006, the unit collected 532 snares, 2 shotguns, 0.5 tusk, 0.1 automatic weapons and 0.1 high-caliber rifles.

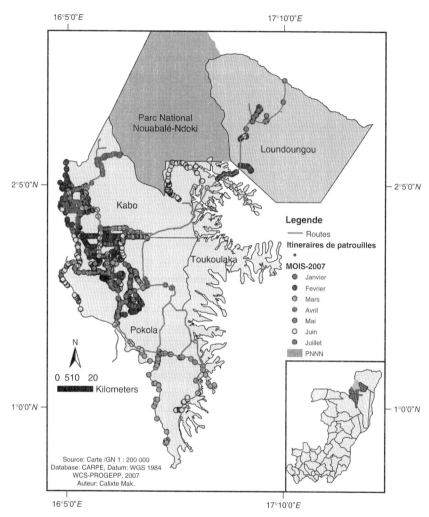

Figure 4.6 **Map of the eco-guard forest patrols conducted over six months in the CIB concessions in 2007.**

Table 4.3 **Results of law enforcement activities for each semester (S1: January–June; S2: July–December) from 2004 through 2006.**

Items seized/ infractions	2004 S1	2004 S2	2005 S1	2005 S2	2006 S1	2006 S2
Wire snares	7062	15,979	10,489	8595	4198	9974
Elephant tusks	9	12	3	10	10	8
Shotguns	43	34	11	36	52	52
High caliber rifles	2	3	1	1	0	1
Automatic weapons	0	2	3	1	2	2
Failure of logging trucks to stop at roadside posts	7	9	14	12	14	9

Evaluating law enforcement results

Results of law enforcement monitoring are difficult to interpret. What does it mean when eco-guards collect more snares or seize more bushmeat during one time period compared to another? Does that mean that eco-guards are working harder and that law enforcement strategies are more effective? Or is illegal hunting on the rise? Conversely, what is the interpretation if eco-guards write fewer tickets or seize fewer items in one time period than another? Does that mean that eco-guards and law enforcement strategies are less efficient? Have animal populations been wiped out, so that less hunting is taking place? Or has law enforcement succeeded so that the rate of illegal hunting has been reduced?

Because of the complexity of interpreting law enforcement results, the success or failure of law enforcement should be inferred in combination with data on wildlife and human populations. After all, the goal of an eco-guard unit is not to arrest hunters; the goal is to stop poaching so that animal populations are not significantly reduced or extirpated. The arrest of 15 poachers a month might be a sign of a job well done if animal populations remain at pre-logging levels. On the other hand, if animal populations are declining, the arrest of 15 poachers a month is likely not sufficient and eco-guards are either failing to do their jobs or the eco-guard unit is too small.

Results from several years of law enforcement monitoring demonstrate that increasing the number of eco-guards leads to a significant increase in the

Table 4.4   Correlation between effort (number of eco-guards, patrols and posts per month) and confiscations and failure to stop at posts. All numbers represent r, the correlation coefficient. Tests for the number of eco-guards are based on 84 months of data and 82 degrees of freedom, whereas the patrols and posts are based on 30 months of data and 28 degrees of freedom. Bold correlation coefficients represent statistical significance at $p < 0.05$. (These tests were also conducted as generalized linear models using a Poisson error distribution. The general results were the same, but the correlation results are presented for ease of understanding.)

Type of effort	Snares	Bushmeat	Shotguns	Large-caliber rifles	Ivory	Non-stops
No. eco-guards	**0.41**	0.03	**0.26**	–	−0.07	**−0.30**
No. patrols	**0.64**	−0.14	0.12	–	−0.11	−0.09
No. posts	**0.57**	−0.21	0.09	–	0.01	0.18

collection and confiscation of snares and shotguns and a significant decrease in the number of trucks that fail to stop at roadside posts (Table 4.4). The number of confiscated snares increased significantly, regardless of the type of effort considered. Statistically, variation in the number of snares is easier to detect than the other indicators because of the large numbers in the forest. Unlike apprehending a mobile poacher, tens to hundreds of snares can be fixed at the same location and stay in place until found. Sweeping the forest is likely to result in finding many snares, whereas finding bushmeat depends on being at the right place at the right time.

While a great deal of information can be obtained from law enforcement, the current monitoring system is fairly simple in order to make it possible for the MEF coordinator and brigade chief to manage the process. Whereas the MEF personnel accumulate the data, BZP researchers enter the data into the database, summarize results and produce maps of forest patrols (Figure 4.6). Additional data, however, would make it possible to predict where hunting is likely to take place. By recording GPS points of the locations where confiscations and arrests are made, spatial modeling with satellite maps could be used to predict the type of forest where hunting takes place. In addition, law enforcement data could be overlain on top of maps of wildlife populations to determine whether hunting endangers critical populations or threatened species. Recording the data at the level of the eco-guard team (patrol or post), rather than summarizing the data by month, would aid in determining the optimal number of eco-guard teams and team size. These data should include the number of members per team, number of confiscations and arrests by team and a code for the general location of the team in the forest.

Monitoring of management strategies

The goal of adaptive management is to determine the most effective wildlife management strategies possible in an environment of uncertainty. Adaptive management involves monitoring and evaluating management strategies, testing what works and what does not and modifying decisions based on newly acquired information (Box 4.8). Since the inception of the project, monitoring the effort and results of BZP activities has been a core part of management (e.g., Elkan & Elkan 2005; Elkan et al., 2006; Clark et al., 2009; Poulsen et al., 2009). Although it is tempting to manage by gut instinct, particularly when familiar with an area or when financial resources are tight, the BZP put a premium on gathering data and using the information to guide its conservation actions.

Box 4.8 **Monitoring for adaptive management**

Although a thorough treatment of conservation monitoring is beyond the scope of this chapter, a few basic rules should be kept in mind.

1. Objective: Specify the desired outcome of a project or activity. A good objective is outcome-oriented, measurable, time-limited, specific and practical (e.g., after 3 years, 75% of school children will know all the protected species in Congo).
2. Indicator: Determine a unit of information measured over time that indicates whether the objective is being met. A good indicator is measurable, precise, consistent and sensitive (e.g., scores from semi-annual quizzes in classrooms on the identity of Congo's protected species).
3. Methods: Write out the methods used to collect data for each indicator. Good methods will be repeatable, avoid bias and balance sample size with practicality (e.g., give students 15 minutes to complete a

quiz naming Congo's protected species, administering the quiz in 1 randomly chosen classroom in every school in the 10 villages in the logging concession).

4. Train, test and retrain: Determine who will collect the data and when and how they will do it. Develop a system to verify that the methodology is being followed (e.g., randomly choose a date and classroom in which to accompany the person administering the quiz to make sure he is following the established procedures).

5. Use the data! Data that is collected but never entered into a computer and analyzed is a waste of money and effort.

The BZP monitoring methodologies are detailed in several procedural documents, including Procedures for the Controlled Hunt, Procedures for Monitoring Wildlife, Zoning Procedures for Wildlife Management and Socio-economic Procedures.

There are many good resources for monitoring, including White & Edwards (2000) and several documents on the Wildlife Conservation Society webpage (http://www.wcs.org).

The BZP monitors the effort and results of its awareness raising, alternative activities and law enforcement. For example, BZP educators record the number of students that participate in environmental education classes and they use both oral and written tests to assess conservation knowledge before and after classes. Alternative activity teams note the number of chickens given to villagers and the number of eggs and chickens produced over time. Similarly, the number of hours spent by an eco-guard team in the forest or at roadside posts is recorded with weekly seizures of shotguns and snares. This information is used to evaluate the efficacy of conservation activities over time and in different situations.

To evaluate the success of conservation in the logging concessions, the BZP also monitors a number of indicators related to the threats to natural resources and the status of wildlife populations. Some of these indicators include: (1) the density of large mammals within the concessions; (2) presence of animals in natural forest clearings; (3) number of animals in the local markets; (4) proportion of bushmeat, fish and domestic meat in the diets of concession households; (5) monthly number of tickets written and wire snares seized by eco-guards; and (6) human population of logging concessions.

The challenge to management is to make use of monitoring data. To guide management decisions, data must be compiled in databases, verified for accuracy and analyzed and interpreted in a timely fashion. This process takes time and is dependent upon technicians with the appropriate skills. The chapters that follow will illustrate how these data can inform management.

5

Assessing the Impact of Logging on Biodiversity in the CIB Concessions

Connie J. Clark[1], John R. Poulsen[1],
Richard Malonga[2] and Paul W. Elkan[3]

[1]Nicholas School of the Environment, Duke University,
Durham, NC
[2]Wildlife Conservation Society, Brazzaville, Republic of Congo
[3]WCS Africa Program, International Programs, Wildlife
Conservation Society, Bronx, NY

With less than 7% of Africa's closed canopy forests protected within parks and reserves, the current area of protected rainforest is thought to be insufficient to ensure the future existence of the majority of species (Soule and Sanjayan, 1998; Fagan *et al.*, 2006). One proposal to mitigate biodiversity loss in tropical forests is the integration of production forests into existing conservation strategies (Rice *et al.*, 1997; Chazdon, 1998; Putz *et al.*, 2000; Pearce *et al.*, 2003; Bhagwat *et al.*, 2008; Chazdon *et al.*, 2009). The idea is that, if properly managed, the large size and varied habitats of production forests could complement the existing system of protected areas, enlarging the 'conservation estate' (Putz *et al.*, 2001). Recent years have seen an increase in forestry companies willing to adopt a paradigm of sustainable forest management that, in theory, promotes biodiversity conservation (ITTO, 2005; FSC, 2006; Clark *et al.*, 2009).

However, debate continues over the most basic questions regarding the effect of logging on forest diversity. Despite a multitude of studies, differences in study methods, logging techniques (reduced-impact logging or RIL,

Tropical Forest Conservation and Industry Partnership: An Experience from the Congo Basin, First Edition.
Edited by Connie J. Clark and John R. Poulsen.
© 2012 Wildlife Conservation Society. Published 2012 by John Wiley & Sons, Ltd.

conventional, etc.), environmental characteristics (time since logging, habitat logged) and species-specific responses to logging disturbance (reviewed by Putz *et al.*, 2000) make it difficult to generalize results. For example, studies conducted soon after logging may produce a different result than those conducted years later since they measure the immediate impact of tree extraction and do not account for effects that appear over time (e.g., decreased reproductive success associated with removal of an important resource, differences associated with stages of succession). Similarly, a species response to logging likely depends on the harvest techniques and intensity employed (e.g., RIL versus conventional logging). Consequently, results of studies are often inconsistent; both increased and decreased diversity in response to logging has been reported (Basset *et al.*, 2001; Dunn, 2004; Craig & Roberts, 2005; Azevedo-Ramos *et al.*, 2006; Meijaard *et al.*, 2006; Cleary *et al.*, 2007; Poulsen *et al.*, 2011). Population level studies are equally inconsistent. Chimpanzee populations, for example, have been reported to increase, decrease or show no change following logging (Plumptre and Reynolds, 1994; White, 1994; Hashimoto, 1995; Box 5.1). Such disparities mean that results of existing research are difficult to interpret and almost impossible to translate into cohesive conservation policy and practice.

Box 5.1 **Great apes and mechanized logging in the Kabo concession (David B. Morgan, Wildlife Conservation Society; Crickette M. Sanz, Washington University; Jean-Robert Onononga, Wildlife Conservation Society; Samantha Strindberg, Wildlife Conservation Society)**

Although mechanized logging is often cited as one of the primary causes of faunal decline in tropical Africa, our knowledge of the specific impacts of logging on most mammal species is far from complete – particularly for great apes such gorillas and chimpanzees. Mechanized logging is thought to have diverse short- and long-term impacts on apes, including alteration of dietary habits, disturbance of social interactions and increasing susceptibility to disease. As the majority of apes in Central Africa reside in forests allocated for logging, gaining a better understanding of how disturbance affects wild apes and ways to reduce the impact is one of the most important endeavors for ensuring their long-term preservation.

Here we present preliminary results of a long-term study to evaluate the effects of logging and associated activities on apes in the Kabo concession. Kabo was logged in the 1970s, but has remained virtually undisturbed since that time. We repeatedly surveyed 87 km of systematically spaced transects within a 75 km^2 area adjacent to the southeast portion of the Nouabalé-Ndoki National Park (NNNP). These transects were surveyed for signs of apes and humans on six occasions between 2004 and 2008, before and during the arrival of forestry teams.

In 2004, chimpanzees and gorillas dominated the landscape with few traces of human presence. This rapidly changed with the arrival of forestry teams and a dramatic increase in the encounter rate of human signs (Figure 5.1). Despite the past logging history and return of forestry teams, chimpanzee and gorilla densities in the timber concession were similar to those in the national park and remained stable throughout the timber extraction process. However, apes were impacted by human activity in other ways. Apes became increasingly wary of human contact during the study period. There was a negative relationship between the frequency of contacting apes and increases in hunting and gathering of non-timber forest products in the logging

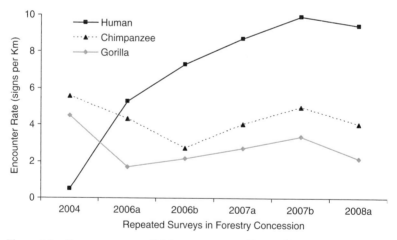

Figure 5.1 Encounter rate of chimpanzees, gorillas and humans in repeated surveys from 2004 to 2008 in the Kabo forestry concession.

zone (Pearson $r = -0.902, n = 6, p < 0.01$). We also found that logging displaced both species at small spatial scales (Morgan, personal communication, 2010). These findings support previous reports that gorillas (Matthews & Matthews, 2004; Arnhem *et al.*, 2008) and chimpanzees (Hashimoto, 1995) seek 'refuge' areas that are less disturbed than forests under exploitation.

Although it is encouraging that density estimates remained stable during logging, we caution that the effects of mechanized logging on apes are likely to be diverse and long-term. For example, the reproductive success of female chimpanzees is thought to have been compromised in forests heavily disturbed by logging (Emery-Thompson *et al.*, 2007) and such changes are unlikely to be immediately detected by population level estimates. Determining the specific impacts of logging on apes will require long-term intensive monitoring of their ecological, behavioral and physiological responses to habitat alteration, encroachment of their ranges and increased competition with other species, including humans. As mechanized logging continues to spread throughout the word's remaining tropical forests, the widespread adoption of reduced-impact logging procedures and certification schemes will become increasingly important tools for conserving biodiversity, including great apes.

To further complicate the issue, the direct effects of logging (alteration of forest structure and increased fragmentation) on biodiversity are often confounded with the indirect effects of logging (hunting, disease transmission and deforestation for agriculture). The opening of forests to logging sets off a domino effect of road construction, immigration of job-seekers and an escalation of agriculture and commercial hunting and trade (Wilkie *et al.*, 2001; Fa *et al.*, 2005; Poulsen *et al.*, 2009). These trends can be particularly devastating for wildlife. In fact, the harvest of wildlife may more strongly contribute to biodiversity loss than the harvest of trees for timber (Milner-Gulland *et al.*, 2003). In the Congo Basin, approximately 60% of vertebrate game species are overharvested (Fa & Peres, 2001).

Given our limited understanding of how logging affects biodiversity, the idea that production forests can meet both forestry and conservation goals is often met with skepticism (Bawa & Seidler, 1998). Only through the establishment and implementation of robust monitoring systems can forestry and conservation managers assess and mitigate the potential impacts of logging on biodiversity. This chapter introduces techniques for monitoring wildlife in multiple-use landscapes. It also summarizes results of monitoring efforts in the Congolaise Industrielle des Bois (CIB) logging concessions initiated to assess the effect of logging operations on wildlife.

To this end, this chapter presents results of various wildlife studies conducted by the Buffer Zone Project (BZP) and independent researchers between 2000 and 2007. It first explores the impact of logging on wildlife populations with data from a concession-wide survey conducted during 2001–2002. The direct impact of logging on wildlife is examined by comparing abundances of large mammals in logged and unlogged forest and the direct and indirect effects of logging on animal populations are decoupled by relating animal observations to habitat, logging history, roads and hunting pressure. The success of the BZP model for conserving populations of endangered and wide-ranging species is evaluated by comparing results of the 2002 survey to those of a second survey conducted in 2006. To improve our understanding of species-specific responses to logging, we present case studies of independent research on buffalo, great apes, duikers, crocodiles and functional groups (such seed dispersers and herbivores). Finally, management recommendations for the conservation of wildlife in tropical logging concessions are offered.

Surveying wildlife in multi-use landscapes

The studies described in this chapter were conducted in the Nouabalé-Ndoki National Park (NNNP) and four logging concessions (Kabo, Pokola, Loundoungou and Toukoulaka) adjacent to the Sangha Tri-national Network of protected areas. These tropical forests are rich in flora and fauna, include several different forest types (Box 5.2) and are characterized by large natural clearings that provide water and minerals to animals (Box 5.3). Logging history in the area extends back over four decades to the late 1960s. When CIB gained concessionary rights to the area in 1997, the concessions included unlogged and logged forest areas. At the time of the concession-wide survey, none of the forests had been logged a second time although CIB started a second logging rotation based on a 30-year cycle in 2006 (Congolaise Industrielle des Bois, 2006). Similar logging techniques were used among years and concessions; timber extraction was low in intensity ($<$2.5 trees/ha) and selective in nature (Table 5.1; Congolaise Industrielle des Bois, 2006).

Box 5.2 **Vascular plant diversity in northern Congo (David J. Harris, Royal Botanic Garden Edinburgh)**

There are approximately 1500 species of ferns, gymnosperms and flowering plants in the BZP area. These range from tall forest trees to tiny saprophytic orchids. The diversity on a global scale is high, but does not reach the levels of the higher rainfall areas of Atlantic forest in Cameroon and Gabon. The number of restricted endemics is low; this is part of the bigger phytogeographical picture. At a continental scale, the semi-evergreen forest occurring across the northwest of the Congo Basin is rather homogenous with many wide-ranging species and few restricted endemics. The conservation implication is that small reserves to protect areas of rare plants are not a high priority; instead the need is to preserve large areas of a functioning ecosystem.

From a continental perspective, the vegetation appears quite homogenous: the whole area is covered in forest and it is very flat with practically no relief. However, five different forest types occur in the area including two types of *terra firma* forest and three types of flooded forest (Harris, 2002; Malanda *et al.*, 2005). The forests on dry land are heterogeneous 'mixed species forest' and homogeneous '*Gilbertiodendron dewevrei* forest'.

Mixed-species *terra firma* forest covers the majority of dry land. The mixed-species forest varies in the number of individuals and species that are deciduous. Forest structure is also highly variable, canopy is not continuous in most places and the forest appears to have a high species turnover. The heterogeneity of the forest is reflected in the observation that rarely does any tree species above 10 cm in diameter at breast height exceed 10% of stem density. The lack of dominance of any single species coupled with variation in habitat types means that there are over 500 known tree species from the Sangha Tri-national area (Thomas *et al.*, 2004; Harris & Wortley, 2008).

The evergreen *Gilbertiodendron dewevrei* forest has lower species diversity than mixed forest. Several of the species that do occur there are at the edge of their range and poorly represented elsewhere, with certain species occurring well outside their main range. Structurally, it is homogenous with few deciduous species.

Flooded forests vary in the depth and timing of flooding. Seasonally flooded forest, along the Sangha River, is dry underfoot for most of the year, but receives an annual influx of silt-heavy water. Riparian forests of smaller rivers have less silt and more tannins – these blackwater forests differ to some extent in species composition. In permanently flooded areas, two species of *Raphia* form a distinctive vegetation type.

About 30% of vascular plant species are only known from one record in the area (Harris, 2002). The *Ficus* genus is the most speciose with approximately 20 species, but individuals are very rare. It is therefore challenging to survey and conserve these rare species.

Box 5.3 **Forests clearings in the Congo Basin: Ecology, conservation importance and current threats (Thomas Breuer, Wildlife Conservation Society)**

Natural forest clearings (called *bais* by the Aka people) are found throughout the tropical rainforest of the Congo Basin, particularly in and around Nouabalé-Ndoki National Park. *Bais* vary in size (from a few square meters to several hectares) and vegetation. Even so, *bais* can generally be classified into two main groups: those visited by elephants and those frequented by gorillas. Most 'elephant *bais*' are located on dolerite outcrops and seem to be created and maintained by elephants digging for minerals. Others are thought to be large blackwater swamps where sediment and organic material have been deposited by flooding during the wet season. 'Gorilla *bais*' are generally located along small to mid-sized watercourses and are usually wetter than elephant *bais*. They are formed in much the same way as elephant *bais* except the floodplain is flatter and wider, making the resulting organic layer much thicker. The vegetation can be described as a 'floating prairie': a close-woven mat of vegetation floating on water or mud (which is generally not strong enough to support a human). Smaller openings without access to water are called *eyangas*.

Bais play an important ecological and social role for many different animals, including birds, reptiles, amphibians and large mammals. Although the clearings are geomorphological features, large mammals maintain them by digging and browsing and keeping vegetation from growing over them (Turkalo & Fay, 1995). The main attraction for large mammals is the abundant protein- and mineral-rich vegetation and soil; minerals such as sodium, which is rare in the tropical forest environment, occur in very high concentrations in *bais*. The principal visitors to *bais* include forest elephants, western lowland gorillas, forest buffaloes, sitatungas, bongos, Congo-clawless otters and spot-necked otters. Occasionally other mammals such as giant forest hogs, red river hogs, black-and-white colobus, marsh mongooses and agile mangabeys visit clearings. A wide diversity of bird and reptile (including long-snouted crocodiles and pythons) species also visit *bais*.

Bais are open habitats that offer unrivalled opportunities for direct observations of mammals. They are also good hunting ground for poachers: animals are often in high abundance and exposed. Logging impacts *bais* through the construction of roads that provide easy access to them. Although buffer zones were established around *bais* in the CIB concessions, ranging from 250 m around large *bais* to 100 m around small *bais* and 50 m around *eyangas*, these buffer zones may not be large enough to protect the hydrology of the *bais* from contamination (including human diseases) and pollution. Noise from logging activities can be heard over 3 km away; elephants and other large mammals react to this noise, sometimes fleeing from bais. Elephants and other large mammals react to this noise and have been observed to flee from *bais* (Breuer, personal communication, 2010; Mbeli Bai). Some regional experts have proposed a buffer zone of up to 2.5 km around large *bais* to help protect wildlife from the impacts of logging, that is, poachers, pollution and stress (Fichlock and Breuer, personal communication, 2011).

Table 5.1 Description of the CIB logging concessions and survey effort (note that the Toukoulaka concession was incorporated into the Kabo and Loundoungou concessions in 2006 after surveys were completed).

Concession	Area (km^2)	Transect effort (km)	No. transects	Year first logged	Annual average logging intensity (m^3/km^2)	Average annual timber volume (m^3)	Human population	No. villages	Road density (km/km^2)
Kabo	2670	758	161	1968	21.6	133,942	4222	7	1.07
Loundoungou	3900	895	186	2002	23.2	163,427	2693	17	0.20
Pokola	3775	1187	59	1968	24.1	144,901	16,299	13	1.08
Toukoulaka	1625	610	149	1992	28.3	137,455	1357	4	1.72
Total	11,970	3450	749				24,571	41	

Approximately 27,000 people use the timber concession forests for non-timber forest products, including subsistence hunting of wildlife. Although law enforcement efforts prevent much illegal harvest of wild game (Chapter 4), even subsistence hunting can be sufficient to alter the structure and composition of the animal community (Poulsen *et al.*, 2009; Box 5.4). Although the NNNP has never been commercially logged, hunter-gatherer populations have inhabited the region for approximately 40,000 years and evidence of iron smelting sites and burning (Box 5.5) has been found dating as early as 1240 BP (Lanfranchi *et al.*, 1998; Zangato, 1999; Brncic *et al.*, 2009). These historical disturbances provide a heterogeneous matrix of habitats even in the absence of logging. The study site is therefore a contiguous yet heterogeneous landscape of diverse forest types, logging histories (time since logging) and human population densities (Table 5.1), offering a unique opportunity to evaluate how even subtle alterations in the biotic and abiotic environment associated with logging and hunting influence wildlife populations.

Box 5.4 **Effects of logging and hunting on guilds of tropical animals (John R. Poulsen, Duke University)**

In tropical forest, logging is nearly always accompanied by hunting, making it difficult to resolve basic management problems, such as whether secondary and degraded forest can sustain ecologically functional populations of tropical animals. We surveyed animals in the Kabo concession and Nouabalé-Ndoki National Park to decouple the effects of selective logging and hunting on densities of forest animal guilds, including apes, duikers, monkeys, elephant, pigs, squirrels and large frugivorous and insectivorous birds.

Animals were surveyed every two weeks along 30 semi-permanent transects positioned in forest disturbed by logging and hunting, logging alone and neither logging nor hunting. After two years, 47,179 animals of 19 species and 8 guilds were observed in 1154 passages (2861 km) over the transects. Species densities varied by as much as 480% among forest types, demonstrating the significant effects of human disturbance on populations of some species. Densities of animal guilds varied more strongly with disturbance than with variations in forest structure, canopy cover and fruit abundance. Independently, logging and hunting had strong negative impacts on some guilds and positive impacts on others: densities varied from 44% lower (pigs) to 90% higher (insectivorous birds) between logged and unlogged forests, and from 61% lower (ape) to 77% higher (frugivorous bird) between hunted and unhunted forests. Their combined impacts exacerbated decreases in populations of some guilds (ape, duiker, monkey and pig), but seemed to counteract one another for other guilds (squirrels, insectivorous and frugivorous birds). Together, logging and hunting shifted the relative abundance of the animal community away from large mammals towards squirrels and birds. Logged forest, even in the absence of hunting, does not maintain similar population sizes as unlogged forest for most animal guilds.

Large-scale, one-off transect surveys are invaluable for measuring densities of the largest mammals, such as elephants and great apes, across millions of hectares of forest (Blake *et al.*, 2007; Clark *et al.*, 2009; Stokes *et al.*, 2010). But smaller-scale, continous sampling designs can be particularly powerful for resolving specific management questions for a fuller suite of animal species (Poulsen *et al.*, 2011). Repeated sampling increases statistical power to detect differences in animal densities across habitats or management units. The results of this study suggest that large, endangered species may not always be good indicators of the health of the rest of the animal community. Elephants are a good example. Elephant densities were not affected by hunting, probably because of law enforcement focused on the protection of endangered species. Elephant densities were higher in logged than unlogged forest, probably due to the higher abundance of saplings and secondary species. Detecting changes in small- and medium-bodied species is important as they serve important ecological functions (seed dispersal, predation, herbivory) and, as game species, they also provide protein and revenue for rural-based communities.

Box 5.5 **Vegetation history of northern Congo (Terry Brncic, Oxford University)**

What impact have prehistoric people had on the rainforests of northern Congo? How has the forest responded to climate changes in the past? How resilient will the lowland forests of the Congo Basin be to future climate change and human disturbances? These are some of the questions addressed by the paleoecological research carried out in northern Congo. The long-term vegetation and climate history of the Nouabalé-Ndoki forest area has been reconstructed for the last 3300 years from pollen, charcoal and geochemical analysis of multiple sediment cores. Changes in relative pollen percentages over time were used to determine changes in vegetation at a local scale. Charcoal was used as a proxy for human-related forest disturbance from burning, and inorganic geochemical concentrations were used to measure dust input into the basin as a proxy for changes in rainfall. Synchronous changes in these proxies were compared over time to determine the relative impact of humans and climate change on Congo Basin forests in the past.

The key findings from these studies indicate that both rainfall and temperature were variable in the Congo Basin during the Late Holocene. Changes in dust concentration show that there were periods with drier conditions than today, such as between approximately 3000–2000 cal yr BP (calibrated years before present) and between 1200–800 cal yr BP. Other periods were sometimes wetter than today, such as between 1700–1200 cal yr BP and 700–150 cal yr BP. The percentages of two montane pollen types, *Podocarpus* and *Myrica*, expanded at Mbeli River during 1700–1000 cal BP. This increase in montane taxa at low elevations suggests that there was a decrease in temperature during this time.

Remains of pottery found along the Sangha River provide evidence that the area has been more or less continuously populated for the last 2300 years. However, microscopic charcoal levels increased dramatically after approximately 1000 cal yr BP at multiple sites. This suggests either an increase in population or a change in agricultural practices such as the use of slash-and-burn.

Pollen results show that central equatorial Africa has been resilient in the face of past climatic and human disturbance events. Moist semi-evergreen forest taxa persisted throughout the last 3300 years with no sign of savanna expansion. However, the forest composition has changed over time. During drier periods, especially during the Medieval Warm Period during 1200–800 cal yr BP, light-demanding taxa and pioneers became more abundant. Wetter periods usually showed an increase in mature forest tree taxa. However, after approximately 700 cal yr BP when the climate became more humid, the percentages of light-demanders stayed high at some sites, corresponding to elevated charcoal levels. The increase in burning seems to have had a bigger impact on forest composition than climate during the last millennium. The high proportion of light-demanding or late-secondary taxa in the forest today may have established after relatively recent forest clearance and may indicate that the forest is still in a state of succession. Contrarily, some of the older mature forest trees may have established under more humid climate conditions 700–150 years ago. These same species may not regenerate as readily in today's drier climate. Given its resilience in the past, the forest may cope with a future reduction in precipitation. It will however be highly vulnerable to degradation if a drier climate is accompanied by fragmentation and burning.

Impact of logging on wildlife: 2002 survey

CIB and WCS collaborated to survey the five timber concessions for large mammal species in 2001 and 2002 (Box 5.6). In addition to examining the impact of previous logging operations on wildlife populations, the survey launched a long-term monitoring system by establishing a baseline for animal populations to which future surveys could be compared (PROGEPP, 2005c). Wildlife surveys were conducted along transects established for a concurrent inventory of timber species, an example of how the BZP partnership collaborated to achieve goals of mutual importance. Coupling CIB tree inventories with wildlife surveys allowed us to increase the total sampling area surveyed and decrease the need to cut additional survey lines (thus increasing precision of the survey while decreasing the labor and financial costs; see Box 5.7).

Box 5.6 **Methods for assessing the status of animal populations (adapted from White & Edwards, 2000)**

Establishment of a wildlife-monitoring program is essential for understanding how wildlife populations are effected by logging, hunting and conservation activities. We provide a guide to the application of line-transect methods (making observations from a straight line traced through the forest) for animal surveys and describe how to collect complementary information on vegetation and human signs, which will help in interpretation of results. Line transects are the most efficient survey method for tropical forests. Our emphasis is on large mammals, but the techniques described are equally appropriate for other taxonomic groups.

Sampling Design and Transect Stratification

Before you begin, think carefully about how you may be able to stratify your study area. Consider the features most likely to affect animal densities, distribution and seasonal movements: (1) locations of human settlements; (2) areas of or adjacent to extractive activities; (3) existing or past roads or trails; and (4) major habitat types (e.g., monodominant stands, mixed forests, extensive swamps, *bais*). As a rule of thumb you should aim to: (1) stratify according to human density; (2) sample each vegetation type in each stratum; (3) ensure good geographic coverage by sampling in all parts of the area of interest; and (4) sample each stratum in proportion to the expected number of animals within it.

Once a (preliminary) stratification has been undertaken, transects should be located from a randomly selected starting point within each stratum and should cross the stratum. They should be oriented to cross major drainage features in order to sample a representative proportion of all vegetation types. The software program 'Distance' has tools to set up a survey plan (http://www.ruwpa.st-and.ac.uk/distance/distancedownload.html).

Cutting transects requires two people, one of whom can use a sighting compass. One assistant traces a path away from the compass person on the bearing, cutting a minimum of vegetation such that it is easy to see which route they took. Cutters must be carefully monitored so that they do not deviate from the path. A straight transect is crucial for good data.

Permanent versus One-off Transects

When planning the transect sampling design you should decide between a one-off (transect used only once) versus a permanent (transect used repeatedly over time) survey design. As a rule, animal species that require direct counts (visual or auditory detection) such as monkeys or hornbills, are best surveyed with permanent transects in restricted areas. If the focal species can be surveyed by counting animal signs (dung or nests) or if you need to survey a large spatial area (e.g., the 1.3 million ha of CIB forestry concessions), one-off transects are appropriate. For many research questions, it is worthwhile adopting both strategies.

Sample Size and Transect Length

Try to obtain a minimum of 60–80 encounters on each transect for density estimation. A short pilot study can be used to assess what length of transect is likely to give you this minimum. There is a tradeoff between increasing the number of transects and increasing the number of encounters per transect. Increasing the number of transects improves the accuracy of your survey. Increasing encounters per transect (requiring longer transects) decreases the variance of your density estimate (improves precision).

Selecting a Reliable Transect Team

The collection of reliable data is essential to conducting a good survey. A researcher or team of researchers must be able to: (1) reliably identify animals by sight and sound; (2) reliably identify animal signs; (3) identify habitat types; (4) measure or estimate distances accurately; (5) record information reliably; and (6) do all of the above while walking slowly, quietly and carefully over long distances.

The number of people who should walk a transect at any one time depends on whether permanent or one-off transects are being used, the number of species being censused and the other information being collected. Permanent transects, once established, are best walked by one or two observers. One-off transects should proceed with a two-phase process: (1) cutters create the transect and (2) the data collection team passes as long after the cutters as logistically feasible (preferably a few days later when the disturbance caused by transect cutting has passed).

Timing of Data Collection

It is best to walk transects when animals are at their most active. Monkeys, birds and duikers are best sampled in the early morning or late afternoon. When counting animal signs only, time of day is not important. Transects walked in different seasons can give different results, particularly for species that migrate from one area to another in search of food or water. If you want to compare results between different areas or habitat types, the transects in all the areas should be walked during the same season.

Information to Collect

In line transect sampling, the observer travels along a line recording detected objects and the distance from the line to each object. For any given animal species, several types of detection may be possible. For most species, however, we suggest recording all direct observations of groups (visual and auditory), nests (for great apes), dung (ungulates) and foot and knuckle prints. For each observation, record the location along the transect and the perpendicular distance.

General Rules for Walking Transects

(1) Transects should be walked at a speed of 1 km/hour for sighting censuses and about half this for dung and nest counts. (2) Stop to listen for animals every 50–100 m. (3) Do not smoke or talk loudly. (4) Do not leave the trail to follow any animal or to count animals beyond your view. It is acceptable to leave the trail to measure perpendicular distances to animals or sign. If you collect information off the trail, clearly record in your notes that it was collected off the trail. (5) The same animal (or group of animals) must not be recorded twice on the same transect.

Measuring Perpendicular Distance

Perpendicular distance should be measured from the center of the transect line to the center of the dung pile. Record the measurement to the nearest centimeter. Note that when objects are found in groups (e.g., groups of monkeys or flocks of birds), measure from the location along the transect to the middle of the group.

Estimating the Age of a Dung Pile

In order to standardize between multiple observers, use a standardized system to age dung: (1) *fresh*: sometimes still warm, with shiny fatty acid sheen glistening on exterior and strong smell; (2) *recent*: odor present (break the boli); there may be flies, but the fatty acid sheen has disappeared; (3) *old*: overall form still present although boli may be partly or completely broken down into an amorphous mass, no odor; and (4) *very old*: flattened, dispersed, tending to disappear.

Estimating the Density or Abundance of Species

Dung: Because visibility varies in different habitats, the dung encounter rate is of little use for comparing between areas: a difference in the number of dung piles may occur in areas where dung density is the same but visibility differs. Dung encounter rates can be used as an index of abundance for monitoring trends in any given area, but overall density should be calculated to assess the reliability of your measurements. To estimate the density of species using dung piles, the following information is needed: (1) density of the dung piles; (2) decay rate of the dung piles; and (3) defecation rate of the species.

Nests: Gorilla and chimpanzee nest counts are undertaken along line-transects in the same way as dung. The following information must also be recorded for each nest encountered: (1) the estimated age; (2) the perpendicular distance to all nests in the nest group, including any which are not visible from the transect (for each nest note on which side of the transect it fell); (3) nest type (see below); and (4) nest height.

Unlike most of the other observations taken while walking transects, you should leave the transect to get an accurate count of the number of nests found in each group, and to look for dung. Gorilla nests tend to be grouped quite closely, although there are sometimes outliers. Chimpanzee nests are generally more spread out than gorillas (>50 m from each other). Nests in trees are often hard to spot and you should look up from several different angles.

Distance analysis: Distance analysis should be used for estimating population density (see Buckland *et al.*, 2001).

Supplementary Information to Record on One-off Transects

By collecting information on vegetation changes or human activity, differences between areas in animal density can be explained.

Human Activity

Observations of human sign from transects should be put in a socio-economic context for the entire region surveyed. For each encounter of human sign, the type of evidence, the estimated age and the distance along the transect should be recorded. Common signs are: roads (used or disused); old village sites; machete cuts or broken stems resulting from a single passage; regularly used human trails; snare lines (active or abandoned); shotgun shell casings; campsites (active or abandoned); fire; current or past agricultural activity; mining activity; large mammal carcasses; and sites of canoe construction.

Habitat Description

A running record of vegetation type, slope, altitude and other variables should be kept. Forest types generally have to be simplified if records taken by different observers are to be comparable. We suggest training researchers to distinguish among the appropriate mix of the following broad categories for your study site: (1) mixed forest with open understory; (2) mixed forest with closed understory (may be similar to old secondary vegetation); (3) mono-dominant forest (note dominant species e.g., *Gilbertiodendron, Garcinia, Berlinia, Lophira, Aucoumea, Julbernardia, Sacoglottis,* etc.); (4) swamps; (5) seasonally inundated forest; (6) coastal scrub; (7) mangroves; (8) Marantaceae forest (note the dominant Marantaceae species); (9) montane or semi-montane forest; (10) liana forest; (11) gallery forest; (12) low closed scrub, often with *Ancistrophyllum* palms present; (13) clearings (elephant *bais* and swamp *bais*); (14) major light gaps; (15) savanna; (16) mature secondary forest; (17) young secondary forest; (18) logged forest (how many years ago?); and (19) plantations (recent or long abandoned; note also the presence of oil palm or mango trees).

Box 5.7 **Methodologies for quantifying the impact of disturbance on wildlife**

Animal surveys based on transects are highly effective at quantifying animal abundance over large spatial scales. Used as presented in this chapter, they can help assess the environmental characteristics that determine the abundance of different species. Moreover, by combining animal surveys with tree inventories, data critical to wildlife management and the assessment of current and future timber stocks can be collected simultaneously, and for less cost. When combining wildlife and tree inventories, a couple of sampling design issues should be kept in mind. First, timber assessments are often carried out along transects across tens of kilometers. To provide statistical power to quantify species abundance, wildlife surveys should consist of many small transects (Box 5.4). Thus, if wildlife surveys are to be conducted on tree inventory lines, it is best to collect data along smaller segments of the line while leaving several kilometers between consecutive segments. Second, if multiple field teams are collecting data, they should be carefully trained with standardized methods. Even so, the transects that they walk should be randomly chosen so that results are not biased by differences in the quality of data taken by different field teams.

Large-scale transect designs are not the only method for monitoring wildlife populations, and the following weaknesses may make them inappropriate for some studies:

1. Data are collected at a single time period, and only represent that point in time.
2. Only species that leave visible signs (e.g., indirect observations of elephant by dung or gorilla by nests) can usually be surveyed because live animals are difficult to detect.
3. The use of indirect observations introduces error into the quantification of animal abundances because the rate of production and degradation of dung or nests is also measured with error.
4. Specific hypotheses (e.g., animal abundance declines with proximity to roads) are often difficult to test because external factors (i.e., hunting pressure) are not controlled.

Some alternatives to large-scale transect designs include semi-permanent transect designs (SPTDs) and before-after-paired-control impact series (BACIPS) designs. SPTDs involve transects that are kept open and walked multiple times. For example, transects could be walked weekly or monthly to quantify the variation in animal densities over seasons and years. Some of the benefits of SPTDs include: (1) the ability to make observations of live animals because the transect is open, only one transect is walked per day during peak animal activity and field teams are small and less perceptible; and (2) potentially greater power to detect changes in species densities due to human disturbances or environmental characteristics (repeatedly surveying the same transect results in greater precision for that transect, but the survey design must still include multiple transects).

BACIPS designs are superior for assessing the impact of a disturbance such as logging and other industrial activities. They can be conducted much like SPTDs except that they include treatment transects that will be affected by the disturbance and control transects that are not affected, and transects are surveyed before and after the disturbance. In this way, changes in animal density from before and after the disturbance are compared between the control and treatment transects to assess the real impact of the disturbance.

Inventory lines separated by 2.5 km were opened from west to east across the concessions. Lines were primarily cut in *terra firma* forest where logging was to be conducted; *Gilbertiodendron* and swamp forest were therefore under-sampled relative to their actual representation[1]. Field teams surveyed for mammals several weeks after the tree inventories were completed. Data were collected along 5 km segments of the inventory lines, leaving an interval of 5 km between transect segments.

Observers walked slowly ($\sim$0.5–0.75 km/hr) along line transects, scanning the ground and forest for animals and their signs. Because signs of animals were much easier to detect than live animals, we present results on dung piles of forest elephant (*Loxodonta africana*), forest buffalo (*Syncerus cafer*), bongo (*Tragelaphus eurycerus*), sitatunga (*Tragelaphus spekeii*) and duiker (blue duiker, *Cephalophus monticola*; medium-sized duikers, *Cephalophus* spp.; and yellow-backed duiker, *Cephalophus silvicultor*) and nests of gorillas (*Gorilla gorilla*) and chimpanzees (*Pan troglodytes*). Direct observations (visual or auditory detection) were used for monkey groups (crowned guenon, *Cercopithecus pogonias*; white-nosed guenon, *C. nictitans*; and gray-cheeked mangabey, *Lophocebus albigena*). Distance along transects was measured to the nearest meter using a hip-chain and topofil, and the distance from the center of each dung pile or ape nest to the transect centerline were measured to the nearest centimeter.

Nest sites were defined as all nests within 50 m of one another that were created by the same ape species and of the same age class. Each nest was classified as definitely chimpanzee or gorilla if verifying signs (direct observation of apes, feces, imprints or hair) were present. As chimpanzees have rarely been observed building ground nests, we attributed nests on the ground to gorillas and arboreal nests of the same age class that were associated with ground nests to gorillas (adopted from methods of Tutin & Fernandez, 1984). Nest age was scored following White & Edwards (2000) for age classes: fresh, recent, old and very old. Field teams encountered and recorded many species, but we focus on those for which we obtained sufficient observations to enable statistical analyses.

---

[1] To estimate average species density in an area with multiple habitat types, densities should be estimated for each habitat and then a weighted average should be calculated based on the relative representation of each habitat in the area.

Changes in large mammal abundance and distribution between 2002 and 2007

To evaluate the success of management initiatives adopted by the BZP partnership, we compare results of the 2002 survey to those of a second survey conducted in 2006. The 2006 survey was a WCS initiative conducted across the entire Ndoki-Likouala landscape (Box 5.8) with the objective of building on previous surveys (Poulsen & Clark, 2004; Blake *et al.*, 2007) to identify changes in the distribution and abundance of threatened species, particularly apes, elephants, bongo and buffalo.

Box 5.8 **Landscape-scale conservation and monitoring of great ape and elephant populations (Emma J. Stokes, Wildlife Conservation Society)**

In 2006, a landscape-wide survey was implemented across the Ndoki-Likouala landscape as part of a long-term monitoring program (Stokes *et al.*, 2010). The survey zone covered 28,000 km^2 of lowland forest comprising two protected areas, the Nouabalé-Ndoki National Park (NNNP) and the Lac Télé Community Reserve (LTCR), and surrounding timber concessions. Of the timber concessions, 13,000 km^2 were under wildlife management and law enforcement programs in the CIB concessions and 2700 km^2 had no formal wildlife management program (Mokabi concession). The primary objective was to assess the population status of great apes and elephants under different management strategies and across different land-use types (Figure 5.2).

Although sections of this landscape had been surveyed previously (Poulsen *et al.*, 2004a, b, c, d; Poulsen and Clark, 2004a; Malanda *et al.*, 2005; Blake *et al.*, 2007; Clark *et al.*, 2009), this was the first systematic survey of the entire zone. The survey used stratified line-transect distance sampling to estimate densities of gorillas, chimpanzees and elephants for each management unit from counts of nests and dung. The sampling effort represented a balance between obtaining sufficient estimator precision and the financial and logistical constraints of conducting

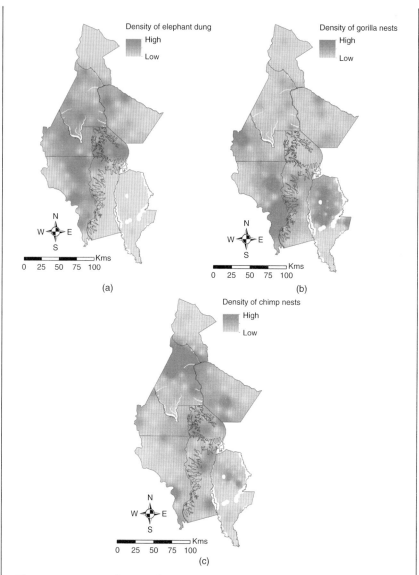

Figure 5.2   **Interpolation of 2006 survey results for (a) elephant abundance from dung density; (b) gorilla abundance from nest density; and (c) chimpanzee abundance from nest density across the Ndoki-Likoula landscape.**

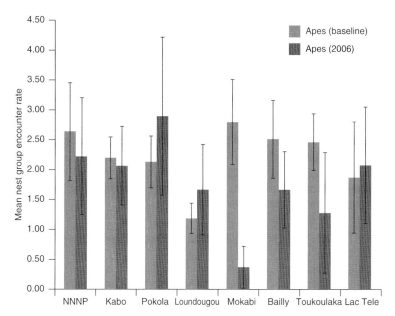

Figure 5.3 **Comparison of great ape nest encounter rates between baseline (2002) and 2006 surveys.**

surveys across such a vast area. A total of 168 transects of 2 km in length were placed systematically across the survey zone (Figure 5.2).

Between February and August 2006, 330 km of transects were walked by 10 survey teams. A total of 829 elephant dung piles were counted. Global elephant density for the landscape was 0.47 individuals/km^2 (95% confidence interval or CI = 0.33 – 0.65), with an estimated elephant population of 13,011 individuals (9305 – 18,195). The highest elephant densities were found in the Pokola concession (0.82 individuals/km^2; 95% CI = 0.47 – 1.42), Kabo concession (0.72 individuals/km^2; 95% CI = 0.46 – 1.13) and NNNP (0.64 individuals/km^2; 95% CI = 0.47 – 0.91). The lowest elephant densities were recorded in the LTCR, with only 5 dung piles recorded on 106 km of line transects (Figure 5.3).

The results confirm that the Ndoki-Likouala landscape harbors one of the largest populations of gorillas, chimpanzees and elephants in Congo Basin. A comparison of our survey results with independent baseline estimates during 2001–2003 (Poulsen *et al.*, 2004a, b, c, d; Blake *et al.*, 2007; Clark *et al.*, 2009) revealed stable populations across timber concessions, with the exception of the Mokabi concession. In the latter case, great ape populations have declined considerably (Figure 5.3) probably due to the lack of an anti-poaching program.

Wildlife management and law enforcement programs implemented in the Kabo and Pokola concessions have likely secured the persistence of elephants and gorillas in these zones. Pokola had the highest overall densities of gorillas and elephants, with gorilla densities surpassing those in the NNNP. For wide-ranging species such as elephants, the data clearly showed the benefits of a landscape management approach for wildlife. The presence of an effective buffer zone management program enabled key elephant migration routes to be maintained across the NNNP, in the Kabo concession to the west and the Bodingo swamps to the southeast (Figure 5.3).

The 2006 surveys were conducted over 2.8 million hectares of contiguous forest representing various management regimes, including two protected areas (Nouabalé-Ndoki National Park, Lac Télé Community Reserve) and six commercially logged concessions (five CIB managed concessions and the Mokabi concession, which has no formal wildlife management strategy). The 2006 survey adopted standard line transect methodology with supplemental surveys *en route* between the transects (White & Edwards, 2000; Stokes *et al.*, 2010).

Estimation of animal density

For the 2002 survey, densities of chimpanzee, gorilla, elephants, buffalo and duikers were estimated in logged and unlogged forest with their associated coefficients of variation and 95% confidence intervals using Distance 5.0, a computer program for designing and analyzing data from wildlife population surveys (Table 5.2; Buckland *et al.*, 2001)[2,3]. See Clark *et al.* (2009) for details of data analysis. Because the number of observations of bongo dung was low and perpendicular distances from transects to monkey groups were not measured, relative abundances for those species were calculated rather than density. Relative abundance is the number of animals or animal signs detected per kilometer walked, whereas density is the number of animals or animal signs detected within a specified area (km^2). Here, the term *abundance* is used to refer to either density or relative abundance.

To compare results of the 2002 and 2006 surveys, both density and relative abundance estimates from the 2006 survey were used (Stokes, 2007).

Direct and indirect effects of logging on wildlife populations

To decouple the direct and indirect effects of logging on animal population observations of each species in the 2002 survey (near the start of law enforcement efforts) were related to (1) habitat, (2) time since logging, (3) distance from forest clearings, (4) distance from primary roads, and (5) human population density[4]. We restricted our analysis to the 2002 data because 2006

---

[2] For robust estimation of detection and the effective strip half-width, furthest observations from the line were truncated (Buckland *et al.*, 2001). Different detection functions were fitted to the data sequentially using half-normal, uniform and hazard rate key functions with cosine, hermite polynomial and simple polynomial adjustment terms. The best model was selected on the basis of the lowest Akaike's Information Criterion (AIC) score and X^2 goodness-of-fit tests were used to examine the fit of the model to the data.

[3] Dung density was converted to elephant density using estimated conversion factors of 19 defecations per day and mean dung lifespan of 45 days for all sites (Barnes *et al.*, 1993). Ape nest density was converted to gorilla and chimpanzee densities using nest decay rates from the Goualougou Triangle, which is adjacent to the Kabo concession, and assuming that weaned chimpanzees and gorillas only make one nest per night (Tutin & Fernandez, 1984; Morgan *et al.*, 2006).

[4] As is common in animal surveys, many transects lacked observations for one or more species - the dataset included too many zeros to meet the assumptions of standard error distributions (Poisson, negative binomial) for modeling animal abundance (Martin *et al.*, 2005). To accommodate this data structure, we used a two-part hurdle model to first estimate species presence and then species abundance given the species is present. We estimated overall mean abundance for each level of the response variable by multiplying the binomial proportion (species presence) by the conditional mean (conditional abundance). We present results as mean species abundance. We calculated mean abundance of each species for each explanatory variable (habitat, time since logging, etc.) using General Additive Models (GAMs), which provide a non-parametric technique for fitting a smooth relationship between two variables (Wood, 2006). Details of the hurdle model and GAMs can be found in Clark *et al.* (2009).

Table 5.2 Abundances of mammal species in logged and unlogged forest in four logging concessions, with their 95% confidence intervals (CI), degrees of freedom (DoF) and coefficients of variation (CV) estimated with Distance 5.0. Abundances were estimated from dung piles for forest elephant, forest buffalo, bongo and duikers (blue duiker, medium-sized duikers and yellow-backed duiker), from nests for gorilla and chimpanzee and from direct observations for monkeys (guenon and mangabey). Note that the measure of abundance (density or relative abundance) differs among species.

Species	Measure of abundance	Forest type	Abundance (95% CI)	DoF	CV (%)
Blue duiker (Cephalophus monticola)	Density of dung	Unlogged	316.6 km^{-2} (162.1, 618.5)	162.1	34.9
		Logged	154.7 km^{-2} (79.1, 302.5)	79.1	35.0
Bongo (Tragelaphus eurycerus)	Rel. abundance of dung	Unlogged	0.003 km^{-1} (−0.04, 0.04)		
		Logged	0.01 km^{-1} (−0.04, 0.06)		
Buffalo (Syncerus cafer)	Density of dung	Unlogged	NA		
		Logged	51.5 km^{-2} (28.6, 92.6)	134.9	30.4
Chimpanzee[a] (Pan troglodytes)	Density of individuals	Unlogged	0.29 km^{-2} (0.23, 0.36)	287.8	10.9
		Logged	0.24 km^{-2} (0.20, 0.28)	560.0	8.6
Elephant[b] (Loxodonta Africana)	Density of individuals	Unlogged	0.38 km^{-2} (0.29, 0.49)	165.9	13.7
		Logged	0.57 km^{-2} (0.47, 0.70)	334.3	10.1
Gorilla[a] (Gorilla gorilla)	Density of individuals	Unlogged	1.92 km^{-2} (1.36, 2.71)	204.6	17.6
		Logged	1.57 km^{-2} (1.27, 1.94)	537.4	11.0

Guenon (Cercopithecus nictitans)	Rel. abundance of groups	Unlogged	0.16 km^{-1} (0.14, 0.18)		
Mangabey (Lophocebus albigena)	Rel. abundance of groups	Logged	0.15 km^{-1} (0.14, 0.16)		
		Unlogged	0.15 km^{-1} (0.13, 0.17)		
		Logged	0.12 km^{-1} (0.11, 0.13)		
Medium duikers[c] (Cephalophus spp.)	Density of dung	Unlogged	1287.0 km^{-2} (986.5, 679.2)	161.2	13.5
		Logged	1706.2 km^{-2} (269.5, 1418.2)	269.5	9.4
YB duiker (Cephalophus silvicultor)	Density of dung	Unlogged	295.3 km^{-2} (211.3, 412.5)	420.8	17.1
		Logged	411.0 km^{-2} (333.9, 505.9)	333.9	10.6

[a] Ape nest density was converted to gorilla and chimpanzee densities using nest decay rates (91.5 days, standard error = 1.67) from the Goualougou Triangle (adjacent to the Kabo concession) and assuming production of one nest per night (Morgan et al., 2006). Average group size was based on the number of nests in fresh and recent nest groups (gorilla = 2.81 nests per group, chimpanzee = 1.68 nests per group). We assume these rates do not vary among sites.

[b] For elephants, we used estimates of 19 defecations per day and dung lifespan of 45 days to translate dung densities into densities of individual animals (Barnes, 2001).

[c] The dung of medium-sized duikers cannot reliably be distinguished by species, so medium duikers includes Cephalophus leucogaster, C. nigrifrons, C. dorsalis, and C. callipygus.

surveys did not sample many of the species included in 2002. Of particular importance are arboreal primates, which we expected to demonstrate a stronger response to the direct impact of logging (e.g., change in the forest canopy) than larger mammals (Chapman *et al.*, 2005). We used ESRI ArcView 3.2 to assign explanatory variables to each transect from existing habitat and logging maps[5].

Comparison of species abundance: Logged versus unlogged forests and 2002 versus 2006

In 2002, transect teams walked a total of 808 transects totaling 3695 km (Table 5.1). For most species, no significant difference in abundance between logged and unlogged forests was identified. Confidence intervals on abundance estimates usually overlapped considerably (Table 5.2). All primates and blue duikers had slightly higher but non-significant abundances in unlogged forest. Elephants and ungulate species had higher but non-significant abundances in logged forest. The lack of observations of buffalo in unlogged forest and the slightly overlapping confidence intervals for elephants suggest that these species may have significantly greater densities in logged forest.

In 2006, observers walked a total of 330 km of line transects and 2314 km of reconnaissance walks. Density and relative abundance estimates from the 2006 survey demonstrate that populations of all species either increased or did not significantly change in the CIB concessions between 2002 and 2006 (Box 5.8). Although not significant (as indicated by overlapping confidence intervals), a downward trend in population size between 2002 and 2006 was observed for chimpanzees and yellow duikers. Despite potential declines for these species, the 2006 survey results suggest that all CIB concessions retained similar or greater abundances of large mammals compared to NNNP. The Kabo and Pokola concessions even demonstrated significantly greater densities of gorillas and elephants than Nouabalé-Ndoki National Park. All five of the CIB concessions had significantly greater abundances of all large mammal species than the Mokabi concession, which was not managed for wildlife (Box 5.8).

---

[5] Habitat categories included: open-canopy forest, closed-canopy forest, *Gilbertiodendron* forest, Marantaceae forest and swamp forest. We calculated the distance (km) from the centroid of each transect to the nearest primary road and nearest forest clearing. Human population density was estimated as the number of people living in camps and villages within 10 km of each transect.

Combined, two lines of evidence strongly indicate that well-managed production forests can enlarge the conservation estate for large mammals: (1) the lack of significant differences in animal abundances between logged forest and unlogged forest (in 2002); and (2) the lack of change in abundances of elephant, gorilla and chimpanzee populations between 2002 and 2006 in the CIB concessions. This is the case at least for the largest and often most threatened species that range over large tracts of forest. Managed forests may be especially valuable when they connect to or surround protected areas within the same forest matrix (Lamb *et al.*, 2005), assuming hunting regulations are enforced (Naughton-Treves & Weber, 2001; Cowlishaw *et al.*, 2005; Elkan *et al.*, 2006).

The effect of logging on wildlife, however, depends on species-specific responses to the logging history and recovery stage of the site, as well as on indirect factors accompanying logging such as proximity to roads and human occupation. Several direct and indirect effects of logging on species-specific abundances are considered in this chapter, and recommendations are provided for wildlife management in tropical production forests.

Logging

Even low levels of timber extraction have been shown to cause high canopy damage (Johns *et al.*, 1996) and edge effects (Laurance, 2000) indicating that reduced-impact logging, although a better option than conventional logging, could still be deleterious for some wildlife species. In this study, we demonstrate that the presence and abundance of species has varied with time since logging, with many species exhibiting non-linear responses to logging (Figure 5.4). In general, abundances of apes and guenons tended to increase with time after logging, nearly doubling after 30 years, whereas mangabey abundance declined. Abundances of duikers either did not change or increased for the first 5–10 years after logging, after which they fell dramatically. Elephant abundance increased for approximately 15 years after logging, then almost returned to its original abundance. Bongo and buffalo abundances did not respond strongly to logging.

These complex non-linear responses to logging suggest that animal abundance varies with different stages of forest regrowth and may explain contradictory results among previous studies. Studies conducted at a single time after logging are essentially snapshots of a population, and therefore ignore a sequence of population level responses to changes in the forest that emerge over time (e.g., decreased reproductive success associated with removal of an important resource). This complex relationship between time since logging and abundance will be further confounded as logging companies begin second and third rotation cycles, a subject that merits further study.

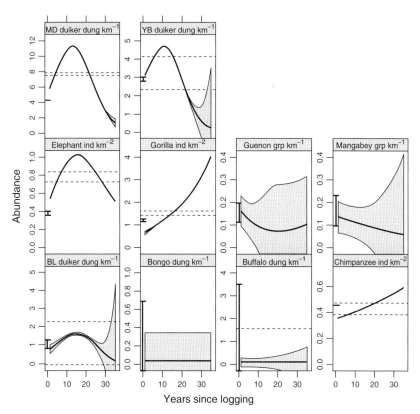

Figure 5.4 **The mean abundance of mammal species as a function of time since logging. Note that units differ among species (ind: individuals, grp: animal groups, and dung: dung piles) as do the measures of abundance (density km^{-2} or relative abundance km^{-1}). Confidence intervals are shaded in gray. The points and bars at year 0 indicate the abundance and 95% confidence intervals of species in unlogged forest. The dashed horizontal lines are 95% confidence intervals for the average mean abundance across all years after logging: if these lines are absent, the confidence intervals for that species are very broad and outside of the plot.**

Changes in animal abundance with time after logging can be explained by changes in forest structure related to the reduction of canopy cover, which generally creates a surge of new growth of grasses, shrubs and herbs on the forest floor (Heydon & Bulloh, 1997; Malcolm & Ray, 2000). For several years after logging, wildlife experiences a net movement of resources to lower levels of the forest strata, which may be particularly beneficial to terrestrial and semi-terrestrial species. Resource gradients shift back towards the upper strata as pioneer trees and seedlings are released from the understory and fill canopy gaps, resulting in decreased density of understory vegetation resources to levels more closely resembling unlogged forests. Vertical shifts in forest resource gradients likely explain the relative decrease in abundances of elephants and duikers 15–20 years after logging, as such shifts can occur relatively quickly (Dunn, 2004). The increase in primate abundance over time is likely due to their ability to exploit a wide breadth of resources from multiple canopy levels (Poulsen *et al.*, 2002) and to adapt to changes in vertical structure when low impact techniques are used. Chapman *et al.* (2005) recorded continued population declines rather than recovery after logging for some primate species, including the gray-cheeked mangabey. Forest specialists with low ecological flexibility, such as gray-cheeked mangabeys, are more vulnerable to habitat disturbance than other species included in this study. In addition to changes in forest structure and resource availability, human activity in previously undisturbed forest can lead to higher parasite loads and transmission of disease between humans and apes (Figure 5.5). The threat of parasites and disease needs to be monitored and mitigated. In sum, temporal responses to logging are complex, often non-linear and depend on the stage of forest recovery.

Habitat and forest clearings

Habitat was less important than other variables for explaining patterns of species presence and abundance. The weak effect of habitat may partially be a result of the study design, which focused on open- and closed-canopy forest where logging occurs: 70% of transects were located in these forest types. Greater sampling in other habitats may have resulted in the detection of larger differences in species abundances among habitat types.

Abundances of elephant (Figure 5.6), blue and medium duikers and (to a lesser extent) bongo and buffalo decreased away from forest clearings, while primate abundance increased with distance from clearings. Elephants

Figure 5.5   Lowland gorillas are still abundant in both the CIB logging concession and the Nouabalé-Ndoki National Park. Here a juvenile gorilla plays on its mother's back in Mbeli bai. Photo by Thomas Breuer.

Figure 5.6   Elephants tussling for dominance in Mbeli bai. Forest clearings are important habitat for elephants, but elephants quickly abandon clearings that are not protected from poaching. Photo by Thomas Breuer.

were strongly influenced by distance from clearings, with their abundance decreasing rapidly to a distance of approximately 15 km.

Our results highlight the potential importance of natural forest clearings (commonly referred to as *bais*; Box 5.3) for duikers, buffalo, bongo and elephants. *Bais* are canopy openings characterized by grass, herbs and browse, believed to attract high numbers of herbivores and browsers by providing locally high abundances of forage, minerals and water (Kreulen, 1985; Ruggiero & Fay, 1994). Such patchily distributed resources may disproportionately influence the movement and spatial distribution of large mammals (Box 5.9; Blake, 2002; Blake & Inkamba-Nkulu, 2004) and should be protected from degradation during logging operations.

Box 5.9 **Survey of forest buffalo in the periphery of the Nouabalé-Ndoki National Park (Richard Malonga, Wildlife Conservation Society)**

The ecology of the forest buffalo (*Syncerus caffer nanus*) is poorly understood despite being the largest bovid in the tropical forests of Central Africa. The forest buffalo may depend more heavily on browse than other buffalo, but an undetermined minimum of grass in their ecological area appears to limit them to the vicinity of grassy glades, river and waterlogged basins (Kingdon, 1997). In a matrix of dense forest, swamp and forest clearing in the northern Congo, Blake (2002) found no evidence of buffalo in primary forest without access to extensive forest clearings. Here we examine the importance of forest clearings on the distribution of forest buffalo in the periphery of NNNP to develop management strategies and recommendations for the long-term sustainability of this species.

Field Methodology

To assess presence or absence of buffalo, we used transects opened by the CIB logging company to survey timber species. For all signs (dung, tracks and observations) of buffalo, we recorded the GPS coordinates and forest type. We used Arc-GIS (ESRI, Redlands California) to estimate distances to the nearest active logging road, forest clearing, human settlement and river, stream or pond. For transects with no observations, habitat was based on the predominant forest type and distances were measured to the midpoint of each transect. Because it was unfeasible to survey for tracks and sign in swamps and forest clearings, all sign detected within 100 m of them were classified as being found within these habitat types. We examined the factors that determined buffalo presence using logistic regression.

Results

Buffalo presence was strongly associated with habitat type. *Terra ferma* mixed forest, the most dominant forest type, covered a total of 47% of the study area; homogenous forest covered 22%; and forest clearings covered <1%. Predicted buffalo occurrence <1 km from forest clearings was more than four times higher in forest clearings than in mixed or homogeneous forest. Buffalo occurrence was high close to forest clearings, water sources and roads and was low close to villages. Closer than 5 km to villages, the predicted buffalo occurrence is nearly 0 for all distances from forest clearing (Figure 5.7). Field data show that buffalo

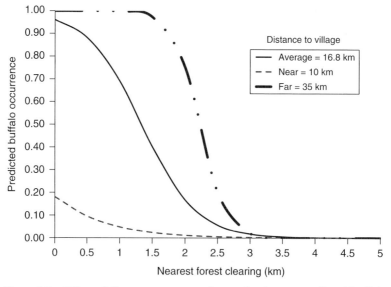

Figure 5.7 **Effect of distance to nearest forest clearing on predicted buffalo occurrence in the CIB concessions. Model estimates control for all other covariates and data are plotted at mean covariate levels.**

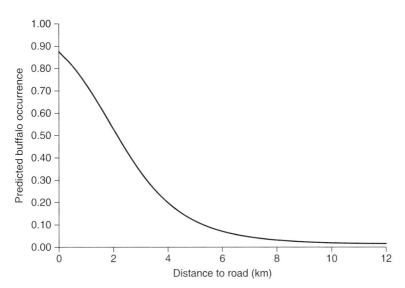

Figure 5.8 **Effects of distance to nearest logging road on predicted buffalo occurrence in the CIB concessions. Model estimates control for all other covariates and data are plotted at mean covariate levels.**

range farther than 3 km from logging roads, with signs present up to about 5–6 km from logging roads (Figure 5.8).

Conclusions and Management Recommendations

Forest clearing and distance to village are the most important factors determining the distribution of buffalo. Forest buffalo distribution was strongly linked to the distribution of forest clearings; even though they represent <1% of the forest area, they provide food, water, minerals and social gathering sites for many large mammals (Elkan, 2003; Ruggiero & Fay, 1994; Turkalo & Fay, 1995). Individual forest clearings occupy small areas and probably do not support viable populations of buffalo. Areas with high concentrations of forest clearings that have the potential to link different habitats should therefore be the priority of conservation. In addition, we suggest maintaining high-quality matrix of forest clearing areas by reducing poaching from nearby roads. No-harvesting buffer zones around forest clearings should also be established to maintain their ecological integrity.

Roads are generally considered deleterious for wildlife because they limit the physical movement of many species (Laurance *et al.*, 2006) and are accompanied by increased hunting, colonization, forest degradation and deforestation (Wilkie *et al.*, 2000; Blake *et al.*, 2007). Species responses to these indirect impacts of logging roads can be obscured by the fact that they occur in concert and may have opposite effects. For example, disturbance of habitat around settlements (e.g., agriculture) could increase resource availability and thus increase the productivity and abundance of some species. At the same time, hunting could decrease the abundance of many of these species, depending on their legal status (i.e., elephants may not be hunted next to villages where poachers could be apprehended). Finally, some species may avoid roads even in the absence of strong hunting pressure. We attempted to decouple these factors by including indexes of hunting, colonization and roads directly into our statistical models.

In the 2002 survey, we found a negative relationship between abundance and proximity to roads for apes and elephants (see also Blake *et al.*, 2007, 2008), but either no relationship or a positive relationship was observed for most other species (Figure 5.9). In fact, abundances of blue and medium duikers decreased strongly with distance from roads. Bongo and the gray-cheeked mangabey also decreased slightly with distance from roads. Roads may provide environmental conditions that some species like: a sunny opening outside the forest canopy or thick roadside vegetation composed of light-loving tree species and herbaceous plants.

To detect potential impacts of hunting on mammal populations, human population density was related to animal abundance as a proxy for hunting pressure. The median number of people within 10 km of transects was 279 (0.89 people/km^2) and the maximum was 7184 (22.88 people/km^2). Of the 22 settlements, one village had a population of over 2000 people; we therefore examined the effect of population on species abundances up to this limit. Most species (with the exception of duikers) were only weakly affected by this relatively low human population, showing little variation in abundance with human population size. Surprisingly, duiker abundances increased rapidly relative to other species over the range of human population densities.

To summarize, medium-sized duikers responded well to habitat disturbance, increasing in abundance in logged areas, along roads and near small villages (where clearing and burning for crops may increase resources). Duiker abundance decreased around villages with over 1000 people, which may be the point at which hunting pressure overrides the positive effects of habitat

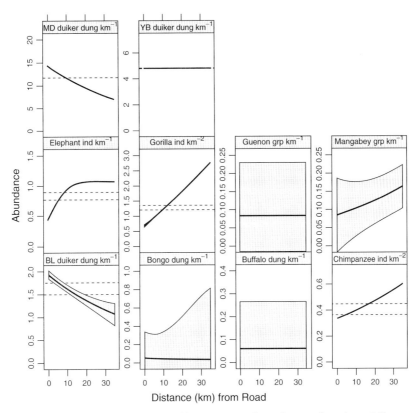

Figure 5.9   **The mean abundance of large mammal species as a function of distance from primary roads. See the legend of Figure 5.1 for more details.**

alteration. A hunting study conducted concurrently with our surveys found no evidence of overharvesting in two villages of <1500 people (Eaton, 2002).

Comparatively, chimpanzees and guenons were negatively affected by population density and roads, but not by hunting. These species may avoid roads even in the absence of hunting pressure along them; alternatively, they may associate roads with hunting and avoid them, but not distinguish different forest areas as threatened by hunting. Elephants, on the other hand, were negatively affected by hunting and roads, but were attracted to villages (probably to raid crops). More research is necessary to determine the types and intensities of human disturbance that affect different species (e.g., Boxes 5.9–5.11).

Box 5.10 **The impact of hunting on duiker home ranges (Miranda Mockrin, Columbia University)**

Duikers (*Cephalophus* spp.) are the most heavily hunted species across forested Central and West Africa. Although they form a vital food resource for people, very little is known about their ecology and demography. The available information was collected from populations protected from human hunting pressure, yet hunting can profoundly alter wildlife demography and behavior (in turn determining the sustainability of hunting). Among ungulates, site fidelity and dispersal behavior over local scales can substantially modify the impacts of harvesting (Novaro *et al.*, 2000). Studies of duiker demography and behavior under hunting pressure are therefore necessary to assess and manage hunting.

In the Kabo logging concession from January 2006 to March 2007, 17 blue duikers (*C. monticola*) were radio-tracked to examine demography and behavior under human hunting pressure. In total, 7 male duikers (5 adults, 2 immature animals) and 10 female duikers (4 adults, 6 immature animals) were fitted with VHF radio collars. Attempts were made to relocate animals every two weeks within a grid divided into 50 m segments. Locations were compiled to calculate home range size and estimate overlap between home ranges for neighboring animals. This study was part of a larger research program examining the sustainability of hunting (Mockrin, 2008).

Home Range Analyses

Twelve animals yielded sufficient relocations to estimate home ranges. For each animal, a 95% kernel density estimator (KDE) was generated, using the *ad hoc* calculation of a smoothing parameter selected by least-squares cross-validation, the current standard in analysis (Kernohan *et al.*, 2001; Thorbjarnarson & Eaton, 2004). Duikers are known to periodically make excursions outside their home range, a pattern also observed in the subjects of this study. The common technique of discarding 5% of outliers, those with the largest harmonic means in the Animal Movement Extension in ArcView 3.2 (Hooge & Eichenlaub, 2000), was therefore used.

Of the 12 duikers, four sets of animals had adjacent home ranges with some degree of overlap in home ranges. The amount of overlap with a neighbor was calculated as:

$$\text{Overlap}_{1, 2} = \text{Area shared between animals 1 and 2}/$$
$$\text{Home range size for animal 1}$$
$$\text{Overlap}_{2, 1} = \text{Area shared between animals 1 and 2}/$$
$$\text{Home range size for animal 2}$$

Results and Discussion

The average blue duiker home range size was 5.86 ha (SD 2.3, Range 3.61–12.10), with no significant difference in home range size by age or sex. Home range sizes in Kabo were similar to those observed in the Democratic Republic of Congo (6 ha) and slightly larger than those observed in Gabon (2.4–4 ha), suggesting that duikers are able to maintain average-sized home ranges under hunting pressure. The percentage of overlap between adjacent animals ranged over 6.5–54% ($\bar{x} = 20\%$, SD 14%) compared to previous studies where adjacent animals did not overlap (Dubost, 1980; Bowland & Perrin, 1995).

It is unclear why home ranges in Kabo would overlap. Home range defense could be reduced if hunting lowers duiker densities and reduces competition for space, or duikers may have changed their activity patterns and behavior in response to human disturbance. Alternatively, activity patterns could have changed in response to an unidentified factor. This study was a first attempt to assess key population parameters and behaviors in a duiker population under hunting pressure. More research will be required to fully assess how hunting pressure alters duiker demography and determines the sustainability of offtake. Only with this knowledge can duiker harvests be sustained, providing a stable food base for human populations while maintaining healthy, functioning forest ecosystems.

Box 5.11 **Crocodilia in Central African forests: A long history of exploitation and the impacts of modern commercial logging (Mitchell J. Eaton, US Geological Survey and Cornell University)**

The majority of effort devoted to research and management of tropical wildlife has been focused on mammals. This is understandable as a large proportion of the biomass consumed by humans, after fish, is derived from mammals. This disproportionate attention suggests a strong bias when considering management of species affected by logging or habitat degradation, however. One non-mammalian group of organisms particularly susceptible to the impacts of logging in tropical forests is the crocodilians.

The forests of northern Congo are home to three species of crocodile: the Nile (*Crocodylus niloticus*), slender-snouted (*Mecistops cataphractus*) and Osborn's dwarf (*Osteolaemus osborni*) crocodile. Exploitation of crocodiles in Central Africa precedes that of industrial logging. Demand from Western nations for leather goods subjected Nile and slender-snouted crocodiles to heavy hunting pressure throughout the 20th century. Residents of the Sangha and Ubangui Rivers still remember the name of the trader who regularly traveled through the region to purchase skins to ship abroad. Hunters from as far away as Senegal came to the Central African coast to hunt crocodiles in the 1970s, extirpating entire populations in only a few months (Eaton, 2006). Because crocodiles are long-lived, slow-growing reptiles – many species only reach reproductive age after 20 years or more – the effects of this harvest on population age structure are still apparent today (Thorbjarnarson & Eaton, 2004). The dwarf crocodile, while not hunted for its skin, is an important food and trade resource for rural inhabitants. In areas of swamp and flooded forest, it constitutes up to 25–30% of non-fish biomass in the bushmeat harvest (Auzel & Wilkie, 2000; Eaton *et al.*, 2009).

Mechanized logging and other commercial enterprises transform subsistence hunting into a commercial venture. The dwarf crocodile is an important target of the commercial and long-distance trade from Central Africa. Its small size and slow metabolism enables it to be captured easily and transported live for several weeks without the need for refrigeration. Vendors often store live dwarf crocodiles to sell when other game become temporarily scarce, making the species an important financial savings instrument. Annually, thousands of live dwarf crocodiles are transported by boat and plane from northern Congo to Brazzaville (Efoakondza, 1993; Thorbjarnarson & Eaton, 2004) where they feed a large urban population or are flown internationally to meet the demand of a growing African expatriate population.

Crocodiles pose unique challenges for conservation and harvest management due to the difficulty of gaining robust inference on demographic parameters in the wild and the magnitude and complexity of the commercial trade. To illustrate our poor understanding of African crocodiles, scientists only recently determined that the dwarf crocodile is not one but a complex of at least three species (Eaton *et al.*, 2009) and that the Congo Nile crocodile is distinct from the East African lineage (Hekkala *et al.*, 2011). New species designations require significant revisions to existing management plans because population size, conservation status and threats to each taxonomic unit may be distinct and require independent consideration. Crocodiles are cryptic and occupy habitats that encumber basic research and monitoring. As keystone aquatic predators, crocodiles play an important role in the ecology of tropical forests. Reliant on freshwater streams in interior forests, they are sensitive to water quality and may serve as ecological indicators of direct impacts to wildlife from stream-flow disruption and sedimentation caused by logging.

Management recommendations and conclusions

This chapter provides evidence that managed production forests can serve to extend, but not replace, the conservation estate for wildlife populations. The results of the studies in this chapter emphasize the importance of monitoring and management at the landscape level. To monitor the full gamut of biodiversity requires large-scale surveys conducted over the duration of the industrial activity. Partnerships such as the BZP can facilitate landscape surveys not only by distributing the cost to multiple partners but, in the case of logging companies, by combining wildlife surveys with timber surveys so that a much larger area can be surveyed.

Similarly, to conserve the full gamut of biodiversity in an area requires protecting a diversity of habitats, including forests logged at different times. The CIB logging concessions, for example, contain several forest clearings that are frequented by large mammals such as elephants. These clearings may partially explain why gorillas and elephants had higher densities in the Kabo and Pokola concessions than the park. In any case, the BZP partnership made wildlife management in these logging concessions possible. As clearly demonstrated by the very low densities of large mammals in the Mokabi concession, management makes all the difference for conservation (Stokes *et al.*, 2010).

Through their incorporation into conservation goals, companies can offset some of the costs of biodiversity protection and can play an important role in wildlife management. The following recommendations are offered to integrate wildlife conservation goals into production forest management and policy: (1) enforce hunting laws; (2) establish hunting and no-hunting zones (in collaboration with indigenous communities) and enforce wildlife laws that limit legally harvested species; (3) establish no-logging and no-hunting zones (conservation set-asides; e.g., Box 5.12) around important and rare wildlife habitat (e.g., natural forest clearings and rivers); (4) close all inactive logging roads to traffic and, if possible, limit all use of active roads to company vehicles and personnel; (5) organize movements of logging vehicles to minimize the periods of vehicle activity on roads and ban driving at night, when most vehicle killings of wildlife species occur; (6) plan logging rotations at long time intervals (e.g., >30 years), allowing sufficient time for vertebrate and plant population recovery while at the same time creating a diversity of habitat within a single concession; (7) implement reduced-impact logging techniques (map and plan tree harvest and road construction, minimize length of roads and skid trails, fell trees directionally to decrease residual damage, etc.); and (8) establish monitoring protocols to identify and mitigate threats to bio-diversity with particular focus on forest specialist guilds with low ecological flexibility (e.g., gray-cheeked mangabey).

Box 5.12 **The importance of set-aside areas for the conservation of great apes in the Ndoki Region: Preservation of the Goualougo and Djeke triangle forests (David B. Morgan, Lincoln Park Zoo, Wildlife Conservation Society; Crickette M. Sanz, Washington University, St Louis; Emma J. Stokes, Wildlife Conservation Society)**

The Nouabalé-Ndoki National Park (NNNP) is internationally known for its important concentrations of endangered species, and particularly large populations of western lowland gorillas (*Gorilla gorilla gorilla*) and central chimpanzees (*Pan troglodytes troglodytes*). The creation of the park in 1993 was only the first step towards the preservation of key ape habitat, however. Beyond the gazetted park boundary, two regions believed critical to ape conservation remained within timber concessions and at risk of exploitation. The Goualougo Triangle, between the Ndoki and Goualougo Rivers, was believed to be one of the least-disturbed areas in all of Central Africa. Indeed, the initial surveys that led to the creation of the NNNP showed the forest to be intact, undisturbed and inhabited by apes that responded as if they had little or no experience with humans (Fay, personal communication, 1999; Morgan & Sanz, 2003). The Djéké Triangle was a second important conservation area for great apes within the logging concession. As of 2010, this area harbors one of only two sites with groups of wild western lowland gorillas habituated to human presence.

Giving its exceptional conservation value, the WCS lobbied for the inclusion of the Goulougo Triangle during the initial planning of the NNNP. This recommendation was not accepted and the long-term protection of the apes remained uncertain. Subsequent discussions among the Congolese government, the WCS and Congolaise Industrielle des Bois (CIB) focused on sparing the intact forests of the Goualougo Triangle from timber exploitation. A flexible land-use planning approach resulted in an agreement that the biological value of the Goualougo Triangle should be maintained and that the forests be formally protected. In 2003, the Congolese government announced that the Goualougo Triangle, comprising 25,000 ha of forest, would be annexed to the park, marking a significant commitment to the long-term preservation of this unique forest and its wildlife inhabits.

Building upon the milestone conservation of the Goualougo Triangle, in November 2004 CIB announced two additional conservation set-aside areas in the Kabo concession as part of its FSC certification process. The two areas, the Djéké Triangle and the Bomassa/Mombongo zone, comprise over 15,000 hectares and are located in the Bomassa Triangle. The Bomassa Triangle provides an important conservation conduit in the Sangha Trinational Protected Area network by connecting national parks in the Central African Republic and Republic of Congo. The Djéké Triangle has never been logged. Both areas contain important complexes of *bais* and *eyangas* (natural clearings frequented by large mammals) and are the subject of long-term ecological research programs. The set-aside agreement recognized the conservation and scientific value of the region and its potential for eco-tourism development. The agreement was reached following stakeholder discussions between CIB, the WCS, the Government of Congo and Stony Brook University Research Group, who first ran the western lowland gorilla research project.

Setting aside the Goualougo and Djéké Triangle forests were landmark conservation initiatives that continue to have far-reaching impacts. Thriving research and eco-tourism projects have been established in these areas. As of 2010, the Goualougo Triangle is the only research site in Central Africa with habituated chimpanzees, and Mondika's gorilla eco-tourism program in the Djéké Triangle is flourishing. Furthermore, important scientific discoveries continue to emerge from these areas. These sites facilitate ape conservation at regional and international levels through education programs and support of Congolese nationals in continuing research and graduate education. Local and national support for these projects continues to grow as stakeholders benefit from meaningful employment and other opportunities associated with these long-term conservation projects.

The conclusions drawn here are from comparisons of forest areas at various stages of logging and recovery. We advocate the before-after-control-impact paired series (BACIPS) study design as the best way to assess the impact of logging activities on biodiversity whenever possible (Box 5.7; e.g., Crome *et al.*, 1996). We recognize, however, that information is urgently needed for management purposes and BACIPS designs are not always feasible as they can take years to complete. Improved understanding of how logging impacts forest diversity could also be gained through studies that: (1) examine entire animal communities and functional guilds, to compare the differential responses among species groups (Box 5.4); (2) correlate animal abundances with fine-scale vegetation and environmental parameters in both logged and unlogged forest (e.g., canopy cover, vegetation density/diversity at multiple canopy strata, changes in resource availability for various feeding guilds proximity to specialized habitats, water courses, roads and hunting pressure); (3) employ identical methods across forests and logging concessions at a regional scale (e.g., Box 5.8); and (4) examine how shifts in animal abundance affect plant regeneration, disease prevalence and meso-predator release. Basic ecological research coupled with monitoring efforts will help scientists and conservation practitioners better understand and mitigate threats to biodiversity as they emerge (Laurance & Peres, 2006).

6

Impact of Industrial Logging on Human Demography and Patterns of Wildlife Harvest and Consumption

John R. Poulsen[1], Connie J. Clark[1], Germain A. Mavah[2] and Paul W. Elkan[3]

[1]Nicholas School of the Environment, Duke University, Durham, NC
[2]School of Natural Resources and Environment, University of Florida, Gainesville, FL and WCS Congo Program, Wildlife Conservation Society, Brazzaville, Republic of Congo
[3]WCS Africa Program, International Programs, Wildlife Conservation Society, Bronx, NY

Overhunting of wildlife for meat across the humid tropics is causing population declines and local extinctions of numerous species (Fa & Peres, 2001; Corlett, 2007; Peres & Palacios, 2007). The statistics are unsettling: 60% of the most common game species in the Congo Basin are hunted unsustainably (Fa *et al.*, 2002), with total harvest of bushmeat in the basin estimated at 1–5 million tons annually (Wilkie & Carpenter, 1999; Fa *et al.*, 2002). The loss of animals and species has consequences for ecological processes that drive forest dynamics (e.g., Wright *et al.*, 2007; Terborgh *et al.*, 2008) and for the rural people who depend on wild meat as a source of protein and revenue (Davies, 2002; Boen-Jones *et al.*, 2003). Overhunting therefore compromises forest regeneration, biodiversity and the livelihoods of forest-dwelling peoples.

Tropical Forest Conservation and Industry Partnership: An Experience from the Congo Basin, First Edition. Edited by Connie J. Clark and John R. Poulsen.
© 2012 Wildlife Conservation Society. Published 2012 by John Wiley & Sons, Ltd.

The drivers of overhunting in the tropics are well known. Expanding road networks fragment the forest, opening it to a growing rural population armed with guns and wire snares (Noss, 1998; Robinson *et al.*, 1999; Wilkie *et al.*, 2000). In frontier forests, where people have no history of commercial agriculture or logging, people exploit the most accessible and abundant resource: wildlife. When roads provide access to markets, bushmeat becomes a market commodity and transforms hunting from a subsistence to a commercial activity. These factors collide and are potentially accelerated by the large-scale operations of extractive industries including mining, oil, industrial agriculture and logging. Fueled partially by emerging economies in China and India, the demand for natural resources and the accompanying rise in prices expand the operations of extractive industries and the pressure on wildlife throughout the tropics (Butler & Laurance, 2008).

In the Congo Basin, logging is currently the dominant extractive industry in terms of number of people it employs and the extent over which it operates. Industrial logging can turn pristine forests into a landscape criss-crossed by roads and dotted with towns (Laporte *et al.*, 2007). Forestry companies pay relatively high wages, thereby growing the local economy and attracting large numbers of people (workers, family members and traders) into areas that were formerly sparsely populated (Wilkie & Carpenter, 1999). The lure of employment transforms logging towns into booming population centers that put unprecedented pressure on forest resources, particularly wild game. Partially because logging takes place in remote forests (i.e., away from urban markets, agriculture and ranching), most companies fail to provide their workers with animal protein and so they survive on bushmeat. Bushmeat is a source of food security and a cornerstone of rural livelihoods. As such, pressure on this limited resource from population growth associated with industry could lead to overharvesting, compromising the ability of indigenous communities to sustain their livelihoods. Logging therefore unites multiple threats to wildlife over large areas; as timber is extracted from the forest, so also is the wildlife.

The town of Pokola in the CIB concessions is a perfect example of the industry-driven transition of frontier forest to population centers. Pokola, located on a crook in the Sangha River, had its beginnings as a fishing and hunting camp. If it was anything like the fishing camps of today, it probably included a couple of one-roomed huts built with rattan frames and stuck together with mud. A wooden canoe or two would be moored to the river's edge. A fire likely burned under a lean-to with a palm-frond roof, smoking the fish to conserve it for later consumption or trade. By the 1970s, Pokola had grown to be a large village in which people had settled to work for the logging industry that was taking root in the region (Auzel & Wilkie, 2000). Pokola acquired its first sawmill in 1986. By 1999 the once-sleepy fishing camp was a bustling town with 8100 inhabitants, including merchants from Cameroon who saw profit in providing goods to logging company workers (Congolaise Industrielle des Bois, 2008). A few years later in 2006, the population of Pokola had grown to over 12,400 people. It was the envy of northern Congo, with full-time electricity and a water supply for every neighborhood. The services and amenities surpassed those of the regional capitols of Ouesso and Impfondo, and included a hospital furnished with medical supplies and equipment, schools, television and radio stations, a marketplace and a butcher shop with freezers to conserve meat (CIB, 2008). Travel to Brazzaville was just a two-hour jaunt by the airplanes that arrived twice weekly.

In addition to infrastructure development, logging provides direct employment to hundreds of people and indirectly employs hundreds more who provide services and goods to company employees. Urbanization and the development of infrastructure and roads are not without their downsides, however. Most of the growth in population comes from immigration. Because industry depends on skilled labor, educated immigrants are more likely to gain jobs than less educated rural people; outsiders therefore accrue the benefits of the forest. The arrival of workers, family members and traders puts additional pressure on natural resources. In addition, as the company employees gain wealth the politics of local communities change, sometimes overturning traditional customs and power structures.

The dramatic changes in Pokola imparted by the arrival of industry have had profound effects upon the social and economic lives of concession residents. This chapter provides a guide to the data necessary to monitor the impact of industrial activities on bushmeat supply, explains how to collect those data and demonstrates how a dramatic increase in human population in the CIB concessions from 2000 to 2006 increased pressure on wildlife resources. It examines trends in human population, the supply and consumption of bushmeat, livelihoods of local people and the impact of logging on the socio-political interactions of local peoples. Although this chapter focuses on bushmeat, the relationship between human demography and other resources could be examined using similar methodologies.

Human population and demographics

The human population of an area is an indicator of the pressure on natural resources. Counting people through a census is the most accurate way to derive estimates of human population. Repeated censuses over time provide information on population change. The BZP has monitored the human population of the CIB logging concessions with annual censuses of all CIB logging towns that account for the majority of the concession residents (Kabo, Pokola, Loundoungou, Ndoki 1 and Ndoki 2; Box 6.1). A census has been conducted in all villages (logging towns and traditional villages) within the concession every five years to get a precise estimate of the total population (PROGEPP, 2005). Censuses provide the opportunity to gather important demographic information such as the number of immigrants and the occupation, education level and age of residents.

Box 6.1 **How to census human population**

The objective of the BZP census was to quantify the human population and demographics in all logging towns. Censusing the logging towns provided a good indication of the level of immigration and population growth in the concessions; traditional villages are comparatively small and so would not have strongly contributed to or detracted from the overall trends.

When conducting a census, the first step is to map and assign a unique number to every house within a town. The map ensures that no households are missed during the census, especially when several people are involved in data collection. Maps can be created by hand or from recent satellite photographs of the town. Once the mapping is completed, the second step is to divide the town into sampling units that can be surveyed in a single workday by a census team. The third step is to visit each house to interview the head of the household. In addition to the number of people in the household, information can be collected on the ethnicity, age, education level, occupation and relationship among household residents.

Importantly, the potential responses to each of these categories need to be determined and standardized before data collection. For example,

---

if a respondent states that he works as a sawyer for the logging company, it must be clear whether this should be recorded as 'sawyer' or 'logging company employee'. The best answer depends on the level of detail that is expected from the census. If potential responses are not standardized beforehand the data will be, at best, extremely messy to summarize and analyze, and at worst, useless.

If resources are not available to census all the towns and houses within a logging concession or industrial area, shortcuts can be taken that still provide good (albeit less precise) answers. For example, if several villages or towns need to be censused, the average number of people per household could be estimated by randomly sampling a proportion of the houses in each village. Multiplying this number by the number of houses in the village would estimate human population. Because people of similar ethnicities and economic class tend to live near each other, the key is to census houses at random from within each of the sampling units.

From a resource management perspective, census data describe the number and identity of potential users of natural resources. In the CIB concessions, local people are supposed to have priority access to natural resources; the original occupants of the forests have a traditional claim to the land and resources. Knowing the population of 'local' people versus immigrants is therefore important for land-use planning (Chapter 3). Similarly, knowledge of the age structure of the population facilitates planning for the future. A population with a high proportion of residents under 20 is a growing population, and long-term planning is necessary to figure out where and how those people will live, hunt and grow food as they becomes adults with children on their own.

Incidentally, these data are as important for industry as they are for conservation. Because CIB is the main builder of infrastructure and supplier of public services, a large human population in its logging towns equals a large investment. The government requires CIB to provide housing for its employees; it therefore needs to know the typical size of a family so that the house design is appropriate. Houses occupied by too many people are also a health and safety hazard; by keeping track of the number and identity of logging town occupants, the company can set appropriate rules to manage town facilities.

Population and demographics of the CIB timber concessions

In the CIB concessions, industrial logging has prompted rapid population growth. In the five logging towns of Kabo, Pokola, Loundoungou, Ndoki 1 and Ndoki 2, the population grew by 69.6% (10,122 to 17,164 people) from 2000 to 2006 (Figure 6.1). This population boom was largely the result of immigration from other parts of Congo: 69% of new logging town residents were migrants, 18% were foreigners from outside of Congo and 13% were local.

Observed changes in population were sometimes the result of logging operations. For example, in 1998 the Ndoki 2 logging camp was established on the east bank of the Ndoki River, creating a population center in a previously unsettled and unexploited forested region. Logging operations expanded into the Loundoungou concession in 2004, creating the Loundoungou logging town. Workers were moved from Ndoki 2 back to Kabo in 2005, although Ndoki 2 remained occupied by settlers who stayed to farm cassava fields (which demonstrates the lasting effect of logging camps as permanent settlements in previously unsettled forest). The growth of Pokola by 5000 people from 2000 to 2006 resulted from a combination of expanded sawmill operations and the immigration of people to take advantage of public services provided by the company (e.g., electricity, water and a hospital).

The inhabitants of the CIB logging towns are young. For example, in the Kabo concession more than half of the population is less than 20 years old, with nearly a third younger than 10 years old. Adults between 20 and 50 years of age represent 42% of the population, and adults over 50 years old comprise just 2% of the population (CIB, 2006). The demographics of logging concessions can be neatly summarized by a statistic from Pokola town: nearly two-thirds of infants less than 10 years old, but only 83 people over the age of 30, were born there (CIB, 2008).

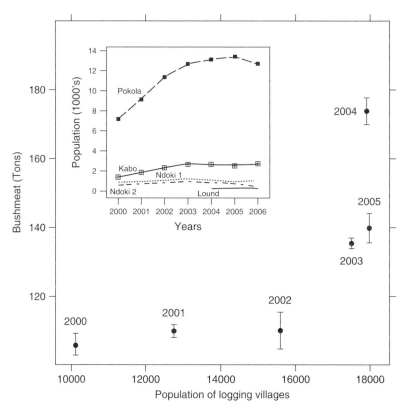

Figure 6.1   Annual biomass of bushmeat entering logging towns versus the combined populations of the towns. Bars are bootstrapped 95% CIs. Inset graph shows human population by logging town from 2000 to 2005.

Supply and consumption of bushmeat

To assess the degree to which changes in human populations influence resources, it is necessary to know either the quantity of the resource in the forest or the use. Ideally, both types of information are collected and synthesized. Estimating the quantity of wildlife was the subject of Chapter 5; this chapter examines the use of wildlife. The supply and consumption of bushmeat are two important indicators of the pressure on wildlife; each tells a slightly different story about the use of the resource. The supply of bushmeat is simply the amount of wild meat reaching markets or households. It provides information on the species composition and amount of bushmeat hunted, but does not tell much about the consumers. The consumption of bushmeat is the amount or type of wild meat consumed by individual households. It provides an estimate of the level of daily protein intake in household diets and an idea of who is fueling the bushmeat market.

These indices are convenient because the data are easily collected (Boxes 6.2 and 6.3); they also have shortcomings, however (Cowlishaw *et al.*, 2005). Data on bushmeat supply are usually gathered where it is most convenient (i.e., at markets or roads). It is therefore often difficult to assess the actual pressure on wildlife. Animals appearing in the market are only a selective proportion of the animals encountered when hunting: hunters may choose to sell large-bodied animals at marketplaces but to eat low-value species in the hunting camp. Similarly, it is difficult to know how much hunted meat arrived by a different route, was taken directly to a town or city outside of the logging concession or was given away or traded within the village without ever reaching the market. Estimates of bushmeat supply will usually be conservative because not all bushmeat is counted. Data on the consumption of bushmeat, on the other hand, usually rely on accounts by people of what they ate. It is relatively easy to determine the frequency of types of bushmeat or other protein sources in the diet, but it is more complicated to assess their importance in terms of mass.

Box 6.2 **How to sample bushmeat offtake**

Bushmeat supply in markets was monitored to determine the factors that influence the amount and species composition of bushmeat in logging towns over seasons and time. To do this, BZP researchers visited bushmeat markets for 10 (Kabo, Ndoki 1, Ndoki 2 and Loundoungou) or 20 (Pokola) randomly selected days each month from 2000 to 2006 (Elkan *et al.*, 2006; PROGEPP, 2005b). At each place they recorded the species of animal, condition (fresh or smoked or whole or part), means of capture (gun, cable trap, spear, net or crossbow), means of transport (canoe, foot, bike or motor vehicle), sale price and ethnicity and principal economic activity of the hunter. Ideally the weight of each carcass or pile of bushmeat would be measured, but merchants were not always willing to let researchers handle the meat. BZP researchers therefore weighed a subset of the carcasses or marketed body part of the animal. The average weight of each carcass or body part was then used to estimate the total biomass (kilograms) arriving in the town.

In most places, not all the bushmeat is sold in markets. In the CIB concessions each town had a single bushmeat market, but a portion of the hunted meat was carried directly to households. Researchers therefore visited markets in the morning when they were most active and before bushmeat had been sold. They also observed the principal trails and roads entering each town for 2 hours in the evening when hunters tended to return from the forest.

Box 6.3 **How to sample bushmeat consumption**

To understand the factors that influence consumption of animal protein (bushmeat, freshwater fish and imported meat) in the logging towns, we conducted consumption surveys for 10 (Kabo, Loundoungou, Ndoki 1 and Ndoki 2) or 20 (Pokola) randomly selected households (new households were randomly selected each month without replacement) on 10 randomly selected days each month from 2000 to 2006 (Moukassa, 2004; PROGEPP, 2005b; Elkan *et al.*, 2006). Research assistants visited households in the afternoon, recording detailed information about the composition of the principal meal of the day including the unit price of animal protein, species of bushmeat, principal economic activity of the family and ethnicity of the head of the household.

People from outside Congo were classified as foreigners, people from a different region of Congo as migrants and people born in northern Congo as indigenous. The term *immigrant* refers to both foreigners and migrants combined. Indigenous peoples include several Bantu groups and Mbendzélé. We also classified bushmeat species into functional groups.

Combining these two indices offers a good picture of how bushmeat is being used in an area. Certain trends in these indices can signal a problem. For example, changes in the composition of species being hunted and eaten can indicate a change in wildlife abundance, perhaps from overhunting (Fa *et al.*, 2002; Crookes *et al.*, 2005; Albrechtsen *et al.*, 2007; Boxes 6.4 and 6.5). Hunting tends to reduce large- and medium-bodied species first, and then target smaller species such as rodents (Fa *et al.*, 2000); an increase in the frequency at which birds and rodents are observed in markets and households relative to duikers and monkeys would therefore suggest that overhunting has caused a shift in animal abundance. Trends detected from data on bushmeat supply and consumption can be confirmed by monitoring animal populations (Chapter 5).

Box 6.4 **Monitoring trends in hunting returns and harvest sustainability: Catch per unit effort (CPUE) (Mitchell J. Eaton, University of Minnesota)**

Catch-per-unit-effort (CPUE) is an index to evaluate changes in wildlife abundance or composition, harvest sustainability and hunter behavior (i.e., de Souza-Mazurek *et al.*, 2000; Rist *et al.*, 2008; Parry *et al.*, 2009). Quantifying CPUE can be a valuable monitoring tool for wildlife management; moreover, CPUE relies on fewer assumptions than market surveys for evaluating the bushmeat supply by assessing hunting offtake and rates directly from the hunter.

In 1999 and 2000, a pilot study was undertaken at Ndoki 2 using CPUE to monitor hunting trends in forestry units surrounding the logging camp. The estimated hunting catchment was divided into three sampling units based on logging activities. These units strongly corresponded to areas of high hunting pressure where hunters used logging vehicles to access the forest. This study took place prior to BZP regulations restricting company vehicles from transporting hunters. Each morning, two to three researchers traveled to active logging areas with trucks carrying forestry employees and hunters. During the study, each logging vehicle carried an average of 7.4 hunters per day. Based on results of earlier trials, we did not accompany hunters into the forest to avoid influencing their behavior and biasing harvest return. Instead, researchers recorded the time and location of entry into the forest for as many hunters as possible. We tried to encounter all hunters as they exited the forest, recording exit time, number and species of animals seen, number and species killed, number of shotgun cartridges carried, number of shots fired and animals wounded but not recovered. Because daily vehicle transport to and from the forest followed a regular work schedule, the number of hours hunters spent in the forest each day was approximately equal across the study period. The unit of effort, therefore, was standardized to one hunter-day. The number of carcasses of each species or taxonomic group and the total biomass killed per 20 hunter-days was analyzed as the unit of catch, which was then scaled to 100 hunter-days for ease of interpretation.

A total of 846 hunter-days and 643 carcasses were recorded during 113 calendar days of hunter follows. Over the two-year study and across three sampling units (totaling 322 km^2), CPUE declined both in the numbers of duikers and primates killed and for total biomass harvested (Figure 6.2). The number of duikers killed per 100 hunter-days declined from 70 to 56 while primate harvest fell from 25 to 13 animals. Total biomass declined at a rate of 11.6 kg for each 100 hunter-days, with hunters recovering an average of 12.2 kg per hunter-day (3.8 kg/km^2) at the beginning of the study but only 7.47 kg per hunt (2.3 kg/km^2) by the end of year 2000. The frequency of hunts resulting in no animal being recovered increased at a rate of 0.31 failures for every 100 hunter-days, reaching an estimated failure rate of 47% at the end of the study from a 34% failure rate at the start.

These findings must be interpreted with caution, however, because initial declines in animal density and availability are to be expected even if hunting is sustainable and harvest mortality is largely compensatory. A decline will continue if hunted populations do not reach a point of demographic equilibrium where replacement is equal to all sources of mortality. Wildlife populations in the Ndoki 2 hunting catchment may not have reached such an equilibrium by the year 2000 which, if harvest rates remained constant, would be considered sustainable even at a lower density. While CPUE trends for Ndoki 2 suggest that the area may be hunted at unsustainable levels, further data collection is necessary to verify continued declines or the stabilization of harvest return for a given level of effort.

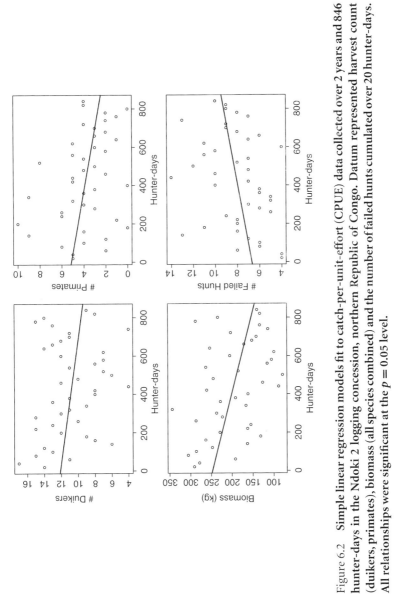

Figure 6.2 Simple linear regression models fit to catch-per-unit-effort (CPUE) data collected over 2 years and 846 hunter-days in the Ndoki 2 logging concession, northern Republic of Congo. Datum represented harvest count (duikers, primates), biomass (all species combined) and the number of failed hunts cumulated over 20 hunter-days. All relationships were significant at the $p = 0.05$ level.

Box 6.5 **Hunting harvest composition and meat division: Links between biological and livelihood sustainability (Michael Riddell, Oxford University Centre for the Environment)**

The Aka (Mbendzélé) hunter-gatherers used traditional hunting methods involving spears, nets, crossbows and traditional snares prior to hunting with guns. Harvested meat was divided among camp and household members depending on their role in the hunt, social position and kinship ties. Large game such as yellow-backed duikers and red river hog provided enough meat to be shared with all households and was consumed over several days. This meat contributed to food security in the forest camp and acted as an essential energy source, ensuring viable forest-based livelihoods.

Today, Mbendzélé hunt with guns provided by their patrons (*nkumu*, see Box 6.7). Meat distribution has evolved with hunting technique, so that it is shared among the gun owner, cartridge owner and hunter. The prey species influences the hunter's reward and the distribution of meat. Large-bodied mammals yield the most meat for Mbendzélé hunters who are often given the head (*mosoko*) and intestines (*mésé* and *mwosé*) (8 kg and 1 kg for the red river hog). If killed in a distant forest camp, the chest bone (*lombo*), heart (*mbombo*), lungs (*mbomboy*), skin (*poso*) and pancreas (*pamba*) are also kept by camp members (~19 kg for a red-river hog). The haunch (*èbèlo*) and shoulders (*obo*) are smoked and carried to the gun owner in the village. If small-bodied duikers and arboreal monkeys are harvested, the head and intestines of these animals are given to the hunter (0.7–0.9 kg for a blue duiker). However, if only one animal is harvested, no meat is rewarded to the hunter. Villagers reported beating Mbendzélé who brought back only small animals.

Between 1998 and 2007/08 the hunting harvest changed significantly. Today, same-day hunts from the village and its environs rarely yield large game. Red river hog comprised 49% of the harvest weight in 1999 and only 33% in 2006, whereas blue duikers jumped from 3.7% of harvest weight in 1999 to 49% in 2006. Duiker meat is low in energy relative to that of hogs meaning that, pound-for-pound, Mbendzélé are getting fewer calories. This causes a physiological strain on individuals. Hunters today also report returning empty-handed more often than pre-2000. Additionally, hunters travel longer distances to find the more profitable large-bodied animals. Before a hunting excursion, villagers give Mbendzélé hunters cassava, salt, cigarettes and flashlights for night hunting. With current harvests, however, the meat from gun hunting lasts only one day in a camp. In the past, meat from small game was sometimes reserved for the hunter's household and not shared around camp due to its limited quantity. Certain cuts of meat originally kept by the Mbendzélé hunter are now given to the gun-owner who insists on receiving this additional meat to sell.

By collecting a few additional data, these studies can go beyond evaluating *what* and *how much* bushmeat was hunted or consumed, and give answers to questions such as *by whom*, *how* and *where* it was obtained (Figure 6.3). To understand the factors that drive bushmeat supply, it could be important to know who is hunting. Are immigrants or local people, company employees or unemployed people doing most of the hunting? Are hunters using traditional weapons, guns or wire snares? Does most hunting take place near the village or

Figure 6.3 Duikers, bushpigs, and monkeys, like the mustached monkey held by a hunter here, are among the preferred prey of local villagers. The Buffer Zone Project monitors both legal and illegal offtake of bushmeat through both market and protein consumption studies. Photo by Michael Riddell.

is hunting being conducted at increasingly greater distances from the village? At what price was the animal sold? This information is useful for assessing whether an area is being overhunted and can aid in enforcement of hunting laws. Furthermore, a sketch of the socio-economic status of hunters and consumers of bushmeat can guide the development of activities to produce alternative sources of protein or income.

Bushmeat supply in the CIB timber concessions

During the period 2000–2005, BZP researchers recorded 29,570 animals ($\sim$345,119 kg of dressed bushmeat) in markets and along trails into the five logging towns (Table 6.1). Two species of duikers were the most common types of bushmeat found in markets (*Cephalophus callipygus* and *C. monticola*), comprising 62.6% of all observations. After duikers, monkeys, bushpigs and reptiles were the most observed species groups. On average, 354 kg of bushmeat arrived in the five towns every day (a total of 129 tons or 95% CI 124.3–133.6 of bushmeat annually). The total biomass of bushmeat was positively related to populations of the five logging towns (Figure 6.1).

In addition to the size of towns, the number of immigrants also greatly increased bushmeat offtake. Immigrants hunted 70% of all bushmeat arriving in all towns, most of which was observed in Kabo and Pokola (Table 6.1). Immigrants also tended to hunt more frequently with snares than guns compared to indigenous people: immigrants snared 30.4% of animals and shot 66.2%, whereas indigenous people snared 5.7% of animals and shot 92.1%. In a comparison of bushmeat arriving in Pokola in 2000 and 2005, the proportion of animals hunted by immigrants was significantly higher in 2005; this suggests that, over time, immigrants became more engaged in the bushmeat trade (Poulsen *et al.*, 2009). Similarly, the proportion of snared animals increased substantially in all towns and the proportion of smoked carcasses increased significantly from 2000 to 2005 in Kabo, Pokola and Ndoki 1.

Of the five logging towns, Pokola was the largest and most developed; it was also the site of the most worrisome trends for wildlife populations. More bushmeat and a higher proportion of smoked meat arrived in Pokola than other logging towns. The prevalence of smoked bushmeat indicates that Pokola has a wide hunting catchment with increased transport times, and that hunters preserve meat to get it to market. The substantial increases from 2000 to 2005 in the proportion of snared animals, animals harvested by immigrants and smoked meat also suggest that hunters are using a larger catchment.

Table 6.1 **Bushmeat data (animals observed during the survey of markets and principal roads) and household meal records.** Bushmeat data are presented as the proportion of all observations killed with different hunting techniques, by hunters of different ethnic original and by condition of the carcass in the market. Household meal data are presented as the proportion of all meals of different types of animal protein and consumed by ethnic origin and principal household. All towns were sampled from 2000 to 2005, except for Loundoungou which was established in 2004. Bushmeat in the meal records came from several species groups: duiker (13,792; 28.0%), monkey (1744; 3.6%), pig (1686; 3.4%), reptile (1221; 2.5%); small- and medium-bodied species (633; 1.3%); insect (534; 1.1%); and large species (188; 0.4%).

	Kabo	Loundoungou	Ndoki 1	Ndoki 2	Pokola	Total
Bushmeat data						
No. animals recorded	2907	1144	4439	2213	18,866	29,570
Months of survey	67	24	69	69	71	
% guns	94	88	87	91	58	70
% other techniques	1	1	3	4	4	3
% wire snares	5	11	10	5	39	26
% indigenes	20	70	56	91	7	28
% migrants	76	27	44	9	57	52
% foreigners	4	3	0	0	36	20
% fresh carcasses	95	99	96	99	64	78
% smoked carcasses	4	1	2	1	35	21
Meal records						
No. of meals	6268	3394	9868	7313	22,077	48,920
Months of survey	66	24	67	66	67	
% with animal protein	91	88	94	93	94	93
% with fish	60	47	47	47	49	49
% with bushmeat	30	47	50	45	41	42
% with domestic meat	3	5	2	5	7	5
% indigenes	25	31	38	23	40	34
% migrants	73	65	61	70	56	62
% foreigners	2	4	1	8	4	4
% logging employees	48	76	38	68	47	53
% hunters	8	9	10	17	3	8
% laborers	10	12	49	12	30	25
% salaried workers	23	3	2	2	3	4
% unemployed	11	0	1	1	17	10

The bushmeat in Pokola included a higher proportion of reptiles and small species (e.g., rodents and birds) than most other towns, and higher proportions of large species including endangered elephants and apes. The greater diversity of species in Pokola markets is likely due in part to the greater use of snares, which are indiscriminate in the species they capture. A single hunter can set hundreds of snares, suggesting that commercialization is the intent of the hunt (Noss, 1998). The human population pressure in Pokola might be sufficiently great that the current level of hunting overtaxes wildlife populations, such that monitoring efforts are detecting the early shift in species composition to smaller-bodied more-diverse assemblages which may be an indicator of over-harvest (Fa et al., 2002; Jerozolimski & Peres, 2003).

Bushmeat consumption in the CIB timber concessions

Records of 48,920 household meals were collected over a period of 6 years. The proportion of meals containing animal protein varied with season and town. Meals in the long rainy season were 30% less likely to contain animal protein than the dry season. The proportion of meals containing animal protein was high, varying between 88% (Loundoungou) and 94% (Pokola and Ndoki 1; Table 6.1). Immigrant and company employee households consumed the majority of meals containing animal protein (Table 6.1). Of meals without animal protein, 75.3% were in indigenous households which lacked animal protein in 15% of meals.

Variation in the type of bushmeat consumed in meals can be either an indicator of preference or of availability of different species or species groups. Meals of indigenous households contained duikers and insects about 1.2 and 4.9 times more frequently than the meals of foreigners and migrants and contained a lower frequency of most other species groups. Households in Pokola more frequently ate small- to medium-bodied species, reptiles and large-bodied species than households in the other towns (with the exception of small- to medium-bodied species in Ndoki 1), but they ate common bushmeat species groups such as duikers and pigs less often. Higher proportions of duikers and insects were consumed during the rainy seasons than in the dry season. Over the six years of monitoring, consumption of duikers, forest pigs, reptiles and large species increased and only the consumption of monkeys decreased.

Livelihoods of people in timber concessions

Hunting contributes to the livelihoods of local people by ensuring access to affordable protein sources and by providing income. While the importance of wild meat for diets and income would seem to promote the sustainable use of wildlife, it depends upon whether alternatives are readily available and whether people perceive them as being readily available.

Bushmeat as a protein source

Bushmeat comprised approximately 39% of the protein in annual diets during the period 2000–2005, reaching as high as 52% when fish was in limited supply. Similar to previous studies (Brashares *et al.*, 2004; Wilkie *et al.*, 2005), the proportion of fish in household diets was negatively correlated with proportions of bushmeat and domestic meat, suggesting that both are likely substitutes for fish. People consumed approximately 20% less bushmeat during the dry season when fish was readily available. We documented a small increase in the consumption of domestic meat from 2000 to 2006. Domestic meat became increasingly available as local traders imported beef and CIB shipped in frozen meat for sale to concession residents. The increase in consumption of domestic meat is a promising sign that people are willing to consume alternative sources of protein.

The presence of bushmeat, fish and domestic meat in meals depended on the ethnic origin and principal economic activity of the head of the household and logging town. The presence of bushmeat was about 20% greater in indigenous households than immigrant or foreign households; when meals of indigenous households contained animal protein, it tended to be wild game. Likewise, meals of people whose principal activity is hunting contained bushmeat more regularly than meals of people with other occupations. Company workers and salaried workers consumed domestic meat (purchased in markets) more than non-salaried people.

Bushmeat as a source of income

Income from bushmeat can be important to the household economy of hunters (de Merode *et al.*, 2004; Cowlishaw *et al.*, 2007). Its importance depends upon how it is used. It is often used as a stopgap for emergencies or to bridge dead times in the agriculture or fishing seasons, but it can also be viewed as incidental income that can be wasted (Solly, 2007). The idea that the forest is immense and animals are inexhaustible can lead to an 'easy come, easy go' attitude towards bushmeat income and decrease the incentive for local people to invest in wildlife management.

In theory, rural households in timber concessions should be less dependent on wildlife for income than areas without employment opportunities. In reality, employment opportunities for unskilled rural communities are often limited and extractive industry benefits skilled immigrant labor (Auzel & Wilkie, 2000). Instead of relaxing pressure on wildlife resources, population growth and the economic boom associated with industry intensify pressure and lead to overhunting. If game species are depleted, marginalized groups and indigenous peoples who cannot easily compensate for the loss of wildlife are especially vulnerable.

Using only the quantity of wild meat arriving in the markets of four logging villages (Pokola, Kabo, Ndoki 1 and Ndoki 2), we multiplied the annual mass of bushmeat in each market by US$0.80 (400 CFA francs); this is a conservative estimate of the price of a kilogram of bushmeat received from a trader in a market and is based on the price of meat of the blue duiker, *Cephalophus monticola*. The average annual return of bushmeat in markets is estimated at US$103,692 (51,846,213 CFA francs) for the four sites over the last 7 years (Table 6.2). Sales of bushmeat represent approximately 1.6% of the cash economy (total of CIB wages[1], PROGEPP wages and bushmeat income) which is divided between the hunters, gun owners and traders who play a role in the bushmeat commodity chain. Although this estimate of revenue is conservative (because it does not take into account bushmeat sold outside markets), even if 50% of bushmeat is sold outside markets the overall sale would only account for 3.2% of the cash economy.

Although bushmeat sales are unlikely to contribute greatly to economic development, they could potentially provide a good source of income to a few local hunters. In addition to being a food source, bushmeat can be an important source of cash income for the rural poor (de Merode *et al.*, 2004). This may be the case in northern Congo, particularly for the Mbendzélé, who are less likely to have permanent jobs with the logging company than Bantus (Lewis, 2002).

---

[1] In 2004, CIB employed nearly 1600 permanent workers and 400 temporary workers, making it the largest logging operation in Congo. It paid approximately US$6 million per year in salaries or about US$3000 per employee (CIB, 2006), well above the average income per capita of US$650.

Table 6.2   Estimates of total biomass (kg) of bushmeat in the logging town markets. The total cost is based on a conservative estimate of 400 CFAfr ($0.75) per kilogram of bushmeat with a conversion rate of 530 CFAfr to US$1.

Year	Kabo bushmeat (kg)	Ndoki 1 bushmeat (kg)	Ndoki 2 bushmeat (kg)	Pokola bushmeat (kg)	Total bushmeat (kg)	Total (US$)
1999	23,496	43,218	35,244	112,528	214,486	161,876
2000	21,716	37,519	27,056	57,494	143,785	108,516
2001	22,072	35,358	16,376	34,653	108,459	81,856
2002	18,156	25,191	18,156	56,946	118,448	89,394
2003	21,360	21,784	13,172	73,560	129,876	98,020
2004	16,020	11,093	11,036	65,949	104,098	78,564
2005	16,376	9317	6764	55,700	88,156	66,533

Social and political organization of local people

In rural villages, such as those found across Central Africa, the level of dependence on natural resource may have encouraged hunters to curb their own use. In such situations, traditional authorities or village institutions often have mechanisms to govern resource exploitation, such as restricting who has the right to exploit them and limiting where exploitation can occur (de Merode *et al.*, 2004). The effectiveness of these mechanisms, however, depends on strong village institutions and interpersonal relationships (Box 6.6; Rose, 2000).

Box 6.6   **Social changes due to forestry and conservation (Michael Riddell, Oxford University Centre for the Environment)**

A number of factors resulting from the expansion of logging and management for conservation have increased the reliance of local peoples on gun hunting. This dependence on guns has also affected the relationships between Mbendzélé men and between Mbendzélé and Bantu. The Mbendzélé have started to abandon the social positions that came with success at traditional hunting; they no longer seek the status brought by spear hunting (*mo.pondi*) when many bushpigs were killed by spear (*bayanji*), or for being celebrated elephant hunters (*tuma*). In addition to the prevalence of gun-hunting, reduced camp size has also contributed to the reduction in net and spear hunting since these activities require a group of individuals to work together.

The relationship between the Mbendzélé and the Bantu has also changed. New relationships with forestry immigrants have been combined with, and in some cases replaced by, traditional relationships. Traditionally, Bantu call Mbendzélé family and they act as household members working without payment; in turn, Mbendzélé call their Bantu family *nkumu*, translating as *my owner* or *patron*. New relationships often involve financial payment for labor but some benefits of the traditional relationship, such as receiving medicine when ill, are not practiced. Approximately half the Makao Mbendzélé households moved to Sombo in 2003 to work for Thanry Congo and have now formed exchange relations with Bantu from DRC, CAR, Congo and Cameroon. Debt relationships are stronger in Makao and Sombo compared to other Motaba villages. Bantu usually initiate these relations and give the Mbendzélé objects such as mosquito nets, pots, pans, clothes and now money in exchange for collecting honey and forest products and working fields. These debt relations now dominate the contemporary livelihood activities of Mbendzélé households.

The arrival of industry impacts the socio-political organizations that govern resource exploitation. Population growth associated with industry puts more people in competition for natural resources. Immigrants have little incentive to use resources in a sustainable way as their future is not tied to the land. Extractive industries can also impose rules and regulations on an area, overturning traditional patterns of management and exploitation.

Industry induces immigration of outsiders who exploit resources and reduce their abundance without regard to local traditions. In northern Congo immigrants hunt ~70% of all bushmeat, demonstrating their significant impact on the natural resources of the area. These same immigrants are disproportionately influential in the management of wildlife resources because they work for the logging company; they therefore have wealth and prestige and are well organized compared to non-workers (Poulsen *et al.*, 2007). CIB employees are organized in unions that lobby the company and government to protect their rights and increase their benefits. In fact, the wildlife and hunting rules adopted by CIB incorporate specific benefits (bi-monthly controlled hunt and certain alternative activities) for workers as a result of negotiations with the workers' union.

Moreover, the relative wealth of company employees also influences patterns of resource distribution, sometimes to the detriment of local people. In the CIB concessions the relatively high salaries of rural workers employed by logging concessionaires provide the means to drive the bushmeat trade (Eves & Ruggiero, 2000; Wilkie *et al.*, 2005). In addition to providing a market for bushmeat, company workers have the means to purchase hunting weapons. They often lend their shotguns to Mbendzélé hunters who lack the cash to buy firearms (Box 6.7). In this exchange the gun owner receives the bulk of the meat, giving the intestines or head of the animal to the Mbendzélé hunter (Box 6.5). Company workers therefore gain cheap meat and benefit from a second source of income by selling a portion of the bushmeat (Wilkie *et al.*, 2001). Without the means to purchase their own weapons, Mbendzélé hunters likely miss an important source of income because it is often the market sale of the meat by the gun owner, and not consumption of wild foods, that can be most important to households living in extreme poverty (de Merode *et al.*, 2004). The high proportion of Mbendzélé meals without animal protein is a demonstration of their poverty compared to other concession residents. It also suggests that the short-term benefits of hunting are being accrued by 'outsiders', to the detriment of indigenous peoples who have prior legitimate claims to bushmeat and other forest resources (Poulsen *et al.*, 2009).

Box 6.7 **Livelihood changes due to forestry and conservation (Michael Riddell, Oxford University Centre for the Environment)**

Makao-Linganga village is located on the eastern bank of the Motaba River, which forms the eastern border of the Loundoungou concession. The population is predominantly composed of Bantu-speaking farmer-fishers and the Aka (Mbendzélé) hunter-gatherers (474 Bantu and 196 Mbendzélé; Mavah, 2005). Both groups originated from CAR and settled in the higher Motaba region in the late 1800s.

Since the early 1990s, a number of developments have led to the availability of salaried work. A conservation project (run by NNNP) was established in 1993 to prevent elephant hunting and to enforce hunting laws. The forestry town Sombo, housing 2500 people, was built just 5 km north of Makao. The Thanry Congo logging company built a road connecting Makao to CAR in 2001. In 2006, CIB constructed a bridge over the Motaba River, effectively joining Makao to the Sangha region. Income derived from these projects has supplemented money earned from traditional forest activities and increased the financial wealth of households, but at the cost of time spent in forest camps. Forest products, including bushmeat, have therefore increased in financial value. Shotgun cartridges have become more affordable to Bantu gun-owners because of their increased income and the cheaper prices that come with lower transportation costs along new roads. Enforcement of hunting laws by the conservation project and social pressure from the Bantu have led Mbendzélé to decrease the use of wire and nylon snares (*waya* and *nylon*) which previously provided over 75% of meat to Mbendzélé forest camps (Kitanishi, 1995) and to increasingly engage in hunting with shotguns.

Due to those factors, Mbendzélé forest camps are now smaller on average (Table 6.3) and located closer to the village. On an annual basis, more time is spent in the village and principal livelihood activities are largely village-based (helping Bantu with household tasks, agriculture, gun-hunting, etc). Correspondingly, Mbendzélé diets now show a higher dependence on agricultural products compared to forest products. With an increase in gun hunting, certain traditional hunting techniques are seldom used and less often found in households, particularly nets (*bo.kia*), spears (*ngongo/ndaba*), crossbow (*mbano*) and traditional snare hunting (*uondo*).

Table 6.3 **Comparison of forest camp characteristics during 1991–1992 and 2007–2008.**

	1991–1992	2007–2008
Mean size of forest camps	46 individuals	11 individuals
Household days in village	52% of days	61% of days
Household days gun-hunting	5%	11%

Extractive industries also change the balance of power at the institutional level. In frontier areas, traditional leaders often govern local communities with little interference from far-away governments. The arrival of industry imposes rules, at least to the extent that local people cannot interfere with resource extraction (e.g., timber or mineral), and is likely to bring greater attention from law enforcement. For example, hunting laws went completely unenforced in northern Congo until the arrival of the timber industry, which made it an issue of international concern. Once the government has leased land to industry, indigenous people have to share lands and resources that they traditionally occupied. They may even compete with the private sector in the establishment of appropriate management and governance systems and the determination of development priorities. Without the financial resources of industry and its easy access to high levels of government, local people are at a disadvantage when it comes to influencing management of natural resources.

Urbanization

Urbanization can intensify the harvest of wildlife in faraway rural areas as it results in a large human population without easy access to natural resources. Where bushmeat markets are booming, rural communities often mine their wildlife resources to subsidize the protein consumption of urban families (Bennett *et al.*, 2007). This can take place inside or outside of timber concessions, but is caused by the failure of development policy (in cities) and timber companies (in logging towns) to provide urban populations with secure livelihoods and sustainable sources of animal protein. It also indirectly impoverishes rural populations who depend on the long-term existence and abundance of wildlife (Box 6.7). In the CIB concessions, the pressure that Pokola exerts on the surrounding forest is evident from the volume of bushmeat that arrives in its markets and the fact that immigrants do most of the hunting using techniques (snares and smoking) designed for commercialization of the resource. Industrial villages such as Pokola are growing up throughout tropical Africa; their impact on forests and forest peoples will depend on the steps taken by government and industry to promote socially and environmentally responsible practices.

Conclusion

Industry can promote biodiversity conservation and human livelihoods by moving toward sustainable practices that explicitly consider the direct and indirect effects of their activities on wildlife (Robinson *et al.*, 1999; Milner-Gulland *et al.*, 2003). The consistency in bushmeat supply over time in the logging towns can be partially attributed to the conservation measures taken by BZP. Many of these measures apply broadly to extractive industry, not just the forestry sector:

1. companies should guarantee the importation or development of protein sources for their workers and their families, keeping prices competitive with bushmeat and fish;
2. companies should contribute to wildlife law enforcement (e.g., salaries of eco-guards who control transport of hunters and bushmeat along logging roads);
3. companies should ensure that their workers hunt legally (with proper licenses and permits) and impose penalties or dismiss workers who break the law;
4. ex-employees should be encouraged to leave the concession to return to their village of origin (an unemployed immigrant is more likely to poach or exploit natural resources than an employed one);
5. traditional systems of resource management (e.g., hunting territories) should be formalized in land-use planning (e.g., management plans for logging concessions) and access to resources for indigenous people should be prioritized;
6. access to forest roads should be restricted to company vehicles, and roads should be closed when not actively used for logging; and
7. urbanization should be avoided in logging concessions (if possible, sawmills and wood-finishing factories should be built and operated in or close to existing cities to avoid the growth of urban centers in the forest).

Although the appropriateness of these measures may differ from site to site, active management for wildlife in logging concessions may be the only way to ensure the persistence of wildlife species in tropical forests.

7

Conclusions and Lessons Learned

John R. Poulsen and Connie J. Clark

Nicholas School of the Environment, Duke University, Durham, NC

The landscape in northern Congo has changed dramatically since 1999. Once limited to a stretch between the Kabo and Pokola villages, roads now form part of the national road system connecting Brazzaville and Bangui. Previously vast virgin forest stands have been inventoried, opened and exploited. A scattering of small villages and fishing camps has become an urban center with amenities that rival regional capitals. The forest has been largely defaunated around Pokola village.

Despite these changes, the available evidence suggests that wildlife populations are still strong. The CIB concessions support densities of large mammals that exceed logging concessions that have not benefited from wildlife management, and are equivalent to densities in the adjacent Nouabalé-Ndoki National Park. At the same time, even though the harvest of bushmeat is considerable and correlated with the size of the human population, there is little evidence that the resource base has been overused. The longevity of these trends is up for question. Are these ephemeral gains? Will future research find that the conclusions from these studies are too sanguine? The answer probably depends on how the BZP partnership develops, evolves and faces the coming challenges.

The very idea that a partnership between a company, a conservation organization and government could benefit forest and biodiversity conservation was radical. The launch of BZP was met with stiff resistance from individuals

Tropical Forest Conservation and Industry Partnership: An Experience from the Congo Basin, First Edition.
Edited by Connie J. Clark and John R. Poulsen.
© 2012 Wildlife Conservation Society. Published 2012 by John Wiley & Sons, Ltd.

and NGOs who accused WCS of selling out and lambasted CIB for white-washing its environmentally destructive activities. Early on, some members of the government questioned the wisdom of extending wildlife conservation to logging concessions, and the administrative procedures had to be developed. But several years later there is evidence that this partnership for conservation was successful in maintaining wildlife populations and biodiversity.

The BZP is an example of a private-sector partnership for conservation (PSPC) and an emerging model for conservation in tropical forests. The relevance of the model is likely to grow given two current and contradictory trends. First, industrial exploitation in tropical developing countries is on the rise, driven by resource-hungry industrial powers (especially China). Second, there is a global recognition that slowing climate change will entail dramatic efforts in forest conservation and stewardship. Approximately 17% of global greenhouse gas emissions are estimated to come from the clearing and degradation of tropical forests (IPCC 2007). This has led the UN Framework Convention on Climate Change (UNFCCC) to try to design mechanisms for compensating developing tropical nations that succeed in reducing carbon emissions from deforestation and degradation (REDD). Thus, even as industry becomes more important in the Congo Basin so does the incentive to actively manage and conserve forests.

Previous chapters described the social and environmental pressures that led to the creation of the BZP. The nature of the partnership between WCS, CIB and the Congolese government and the institutional organization of the BZP have been discussed. The activities undertaken by the project, from land-use zoning to awareness-raising and community conservation to law enforcement and adaptive management, were outlined. The success of these activities was evaluated by looking at results from several years of monitoring animal populations, human populations and bushmeat supply and consumption.

It is our hope that the PSPC model of conservation can be replicated and scaled up so that biodiversity conservation extends beyond protected areas to multiple-use landscapes (Box 7.1). This final chapter synthesizes the primary lessons learned from the BZP. Not every PSPC will be organized like the BZP or include the same activities. However, there are some widely applicable lessons from the project that can facilitate future conservation efforts. The components of the partnership that made the project successful are discussed below, followed by recommendations to improve the BZP and similar partnerships for even greater conservation success.

Box 7.1 **Steps for replicating the BZP model**

The following steps are recommended in the development of a PSPC.

Determine the Conservation Goal

The first step in the establishment of a PSPC is to define the goal of its existence. Over what area will the PSPC work? Will conservation be constrained to the industrial site or include other lands (e.g., adjoining protected areas)? Will conservation goals be broad (landscape conservation) or narrow (specific species)?

Assess the Threats of the Industrial Activity to the Conservation Goal

Before management actions can be implemented to achieve the conservation goal, the impacts of the industrial activity on biodiversity and livelihoods of local residents must be determined. The PSPC would ideally be organized before the initiation of industrial activities, in which case threats should be determined from a review of industrial impacts in other areas.

Identify the Appropriate Partner Organizations for the PSPC

With knowledge of the conservation goal and the potential threats of industry to the goal, the appropriate mix of partners can be identified for the PSPC. The scale of the conservation goal will largely determine the type of partner organization to incorporate. Once organizations have been identified, the partnership should be formalized through a protocol or memorandum of understanding (MOU) that clearly defines the roles and responsibilities of each partner.

Quantify Pre-industry Biodiversity and Livelihood Baselines

If possible, biodiversity and socio-economic livelihood surveys should be conducted before resource extraction occurs. Such surveys could establish a baseline to assess both the impacts of industry and the impacts of management actions taken for conservation.

Plan and Implement the Management Actions to Mitigate the Threats to Biodiversity

Once the planning is complete and the PSPC built, the hard work begins with the definition of conservation strategies and the implementation of management actions. Because multiple drivers of biodiversity loss usually occur simultaneously, a multi-faceted approach that tackles both environmental and social threats will often be the most effective.

Use Adaptive Management to Refine Management Actions

The final step is to re-evaluate conservation strategies and management actions so that successful actions can be continued and failed actions can be revised or discontinued. The monitoring of conservation indicators is therefore necessary for a quantitative and objective evaluation of whether management actions are obtaining their desired results.

Lessons learned: What worked?

Some of the principal components of the success of the BZP partnership are discussed in the following sections.

Conserving landscapes

Strategies to protect tropical forest diversity outside parks and reserves present an opportunity to enlarge the conservation estate. With greater area comes the possibility of increasing connectivity across the landscape, maintaining viable populations of endangered species and conserving genetic variability. By expanding the area under management, PSPC's can buffer core areas (parks and reserves) from encroachment. A single protected area managed in isolation may be too small for the long-term conservation of wide-ranging species; the effort and money invested into species conservation is spent in vain when animals stray across park borders into unmanaged and unprotected lands. For species such as elephants that travel hundreds of miles, the conservation of multi-use landscapes may be their only real hope for survival

(Blake *et al.*, 2007). The BZP extended the protection of wide-ranging endangered species from $400 \, km^2$ to $1700 \, km^2$ by managing hunting and wildlife in concessions adjacent to the NNNP.

Taking a multi-faceted approach to wildlife conservation

The multi-faceted approach to landscape conservation undertaken by BZP has been one of its greatest strengths (Elkan & Elkan, 2005; Elkan *et al.*, 2006). Through the combination of alternative activities, awareness-raising, law enforcement and application of company rules related to hunting and wildlife conservation, both 'carrots' and 'sticks' have been employed to incorporate local people into biodiversity conservation. Before the creation of the BZP, local people exploited wildlife and forest resources without regard to national laws or long-term sustainability. Logging company employees hunted on and off work hours, using company infrastructure for commercial poaching and trade. Providing information to local communities about Congolese wildlife laws and conservation helped them understand the importance of wildlife management and law enforcement, particularly in the context of a growing human population. The alternative activities program offered revenue and protein alternatives to hunting, essentially as a reward for accepting law enforcement. Integrating wildlife regulations into the CIB company rules linked the economic incentive of employment to respect of national wildlife and hunting regulations. This multi-faceted approach has also combined research and monitoring with conservation, facilitating an adaptive approach to management.

Formalizing land-use planning

Land-use planning that designates areas of resource access and exploitation should be informed by participatory processes and formalized through management plans. Management plans typically describe an activity (industrial or otherwise) to take place in a site, the impacts of the activity on the environment and society and the plans for mitigating any negative impacts. Management plans have been written and adopted by the government for the Kabo, Pokola and Loundoungou concessions. These plans, written by CIB in collaboration with the BZP, described in detail the wildlife management system developed and put in place by the BZP. The plans therefore incorporated the CIB company rules on hunting, the demarcation of hunting and no-hunting zones, alternative activities, law enforcement procedures and wildlife monitoring procedures (e.g., CIB, 2006).

In the past, Central African countries have been lax in requiring management plans of companies and conservation organizations. CIB operated for years in northern Congo without a formal management plan. This trend is changing. Even if all stakeholders have participated, negotiated and agreed upon the zoning and rules for exploiting resources, the procedures and principles must be incorporated into a formal plan. This guarantees that the plan is in agreement with national (and sometimes international) laws and standards. It also ensures that outside actors respect the plan. In 2007, for example, the strength of management planning for biodiversity conservation was put to the test. A government official delivered a large-game hunting permit to a group of expatriate hunters and directed them to the Kabo logging concession. According to the hunters, the official told them it was the only timber concession in Congo with abundant animals and easy hunting. Hearing news of a group of European hunters crossing into the concession, the CIB general director called the WCS principal technical advisor to warn him of the problem. With a phone call to the official and a friendly reminder that the Kabo management plan prohibits safari hunting, the mistake was recognized and corrected. The hunters were re-directed to a different forest concession.

By designating spatially explicit zones for different types of extractive activities, the land use plan is the principal method for assigning 'ownership', and thus authority, to control access to an area and its resources. A key lesson that has emerged from this experience is that participatory mapping is an effective method to promote land tenure and empower local people to manage their own natural resources. Thus, even though the government did not previously acknowledge traditional land tenure, village views of tenure have now been incorporated in the officially sanctioned management plans.

Involving multiple actors in land-use planning

Resource extraction zones such as timber concessions generally serve multiple purposes. Most tropical forestry concessions were home to indigenous peoples and habitat for wildlife long before concessionary rights were sold to logging companies. Prior to logging in northern Congo, nearly 12,000 people lived in permanent villages and temporary camps, making their livelihoods from the forest. Timber production should therefore be understood as an additional activity being introduced into an existing landscape of ecological, livelihood, economic and cultural activities. As such, there are more stakeholders than concessionaires with interests in the forests and their future; these interests must be incorporated in the land-use planning process.

To incorporate all actors, there must be a platform by which they can express their needs. In particular, local communities tend to be less empowered than formal organizations such as companies, NGOs and worker unions. By working directly and frequently with local communities over several years, the BZP has helped to promote indigenous people's rights (including conservation of their traditional territories) to the company and the government. During the writing of concession management plans, CIB also held village meetings to involve communities in the process of developing the plans and to obtain their clear, prior and informed consent. After the formal management plan was drafted, village leaders and other local people again had the opportunity to express their opinions, opposition, and needs at open forums before the plan was finalized. In addition to making the land-use planning process as open and transparent as possible, there should also be a mechanism for conflict resolution for situations when stakeholders simply cannot come to agreement.

Setting aside valuable habitat

An important component of land-use planning is setting aside valuable habitat for wildlife and for ecological processes. CIB has removed nearly 30% (450,000 ha of its 1.47 million ha) of its forest concession from timber harvest. The land-use plan for the CIB concessions protects large forest clearings and riparian areas as part of their commitment to Forest Stewardship Council (principles 6 and 9: environmental impact and high conservation value forests). Not only are both of these areas off-limits to logging, but buffer zones limit timber harvesting to a distance of 50 m from them. The most substantial set-asides include areas of known importance for protected species. The Goualougou triangle, a 26,000 ha region on the southern end of the Nouabalé-Ndoki National Park known for its intact fauna and high density of naïve (unhabituated) chimpanzees, was removed from the Kabo concession and eventually annexed to the park in 2001. In 2004, the Djéké triangle and the Mombongo/Bomassa zone were removed from harvest because they housed the world's only group of habituated lowland gorillas and abundant populations of bongo and forest elephant[1].

Controlling road development

Roads open access to forests and link them to markets, they facilitate the colonization of land by subsistence farmers and the commercialization of forest resources. A large body of literature attests to the strong effects of roads on land clearing for agriculture (Kaimowitz et al., 1999; Barreto et al., 2006). Animal abundance is often lower near roads because of the higher incidence of hunting and effects of forest fragmentation (Laurance et al., 2006; Blake et al., 2007; Clark et al., 2009). The negative impact of roads on wildlife can be minimized through good road planning, limiting the surface area of roads and closing unused roads. In the CIB concessions, the width of roads is limited to 15 m and all roads are blockaded once they are no longer needed for logging.

Probably the single best way to reduce pressure on wildlife is to direct roads away from valuable conservation areas. This can be achieved by carefully planning road layout. Within the BZP, road layout has been discussed prior to the construction of new primary roads to identify and avoid key wildlife habitats and populations. Despite these efforts, CIB failed to take into account a key recommendation to build its primary road through the Loundoungou concession at least 30 km from the NNNP. By building a road and logging town just 16 km from the park, the company exposed it to considerable poaching pressure – probably the single greatest threat to species conservation within the park borders.

---

[1] While these set-asides were great achievements, proposals for a larger no-cut zone (e.g., limiting logging to 100–500 m from the park borders (rather than 50 m) were rejected by the MEF and CIB. In 2004, CIB cut several trees inside the southern border of the park, a mistake which could have been avoided if a larger no-cut zone had been established.)

Employing adaptive management based on data and balanced by economic and social needs

While an experienced conservation manager may often make good decisions, data and information are necessary to justify those decisions to the PSPC members and outside critics. Adaptive management is a structured learning process for decision making, with the aim of reducing uncertainty over time through monitoring and evaluation. Often characterized as 'learning by doing', adaptive management must be evidenced-based and rigorously empirical. Data collection is key to the process.

Whenever possible, management decisions taken by the BZP have been based on data collected over many years. Before management plans were written for the CIB concessions, the BZP had completed studies on wildlife populations, bushmeat, NTFPs and timber species in addition to socio-economic studies of the movements of semi-nomadic peoples, their traditional territories and annual demographic censuses of the human populations within the concessions (Auzel & Wilkie 2000; Eves & Ruggiero 2000; Eaton 2002; Elkan *et al.*, 2006; Poulsen *et al.*, 2007; Mockrin, 2008; Clark *et al.*, 2009). These studies helped to identify the drivers of pressures on wildlife, and determined where set-asides should be created and the sizes of buffer zones along roads and forest clearings.

Even with the data to justify conservation and management measures, economic and social considerations often win out. For example, although a 15 km buffer around forest clearings would be best for elephant conservation in the CIB concessions, a buffer of only 500 m was established with the economic justification that larger buffers would considerably reduce the area for timber harvest and the company's ability to meet its financial targets.

Prioritizing the rights of indigenous people to land and resources

Industrial activities often alter traditional patterns of natural resource use and social power, and therefore particular attention must be paid to protecting the rights of local and indigenous people. The rights of indigenous peoples (defined here as people who lived in the area before the arrival of industrialized logging) to the lands, territories and resources they have traditionally occupied include the right to exert control over access to lands, to establish management systems and to maintain cultural and intellectual heritage.

Land-use planning in the forest concessions of northern Congo has included consultation of local communities, but local residents have not yet established their own formal management systems. Adoption of a zoning system based on land-use practices of indigenous people was a first positive step towards reinforcing local authority over their traditional hunting, fishing and gathering zones. At present, however, policy decisions are still largely initiated by the logging company, the government and the BZP. Progress must still be made in preventing management decisions in logging concessions from marginalizing indigenous populations. For example, CIB employees enjoy privileges not extended to non-workers because of their relatively greater wealth, their ability to organize themselves through worker unions and the simple fact that the logging company is interested in treating its workers well so that they are productive. This puts non-workers at a disadvantage because they lack organization and representation. This is particularly true for the Mbendzélé, whose lack of formal education, attachment to an 'immediate return' economy and forest lifestyle results in a lack of representation within the logging company and government (Lewis, 2002).

It is important that *all* partners embrace participatory processes in engagement with stakeholders; it is altogether too common to attribute community leadership to the wrong individuals, and even more common to confuse support from community leaders or elected officials with support from communities.

Promoting certification schemes

The growing market for certified wood, particularly in European countries, has started a trend in better forest management and land-use planning. Many countries require that imported wood come from legal and sustainable sources. Several forestry concessions (including the Kabo and Pokola concessions) have now been certified by the FSC in Central Africa, and several companies have committed to seeking certification in the coming years. Companies only receive certification if their logging procedures meet the standards set by the certifying body, as determined through independent audits. Auditing is a systematic process of verification, usually conducted at the level of the forestry concession, to determine whether the operation meets a predefined set of criteria or performance standards. If operations meet the minimum standards, a certificate is issued. If not, corrective actions may be requested. The corrective actions must be completed within a specified timeframe for certification to be granted. Subsequent spot checks and audits are then conducted to validate the certificate. For producers such as CIB, certification brings systematic management, potential market access and an improved image in the eyes of investors, consumers and regulators. For conservation, certification provides a mechanism for influencing management practices. For consumers, it provides information on the legality and environmental and social impacts of the wood being purchased. To date, the only internationally recognized performance-based scheme issuing certificates for tropical forests is the Forest Stewardship Council.

Certification brings prestige to companies and promises access to new markets and higher prices for timber. In northern Congo, FSC certification motivated CIB to take a more active role in wildlife management and conservation than it previously had. Certification alone, even with its standards and audits, would not have resulted in the wildlife management system developed through BZP however. Logging companies rarely have the expertise or resources to design and implement a comprehensive biodiversity management system. With WCS, CIB acquired a technical partner to develop and implement a wildlife management system. With MEF, CIB acquired a partner with the mandate to manage eco-guards and enforce hunting laws. The lesson from the BZP is that certification can be a motivating factor, but that multi-organizational partnerships are often the necessary tools for achieving biodiversity conservation and wildlife management.

Forest certification schemes fall short when it comes to wildlife management and biodiversity conservation because of a lack of clear objectives, targets and definitions of the scale at which these are measured (Bennett, 2001). For example, although most certification bodies address wildlife conservation to some extent, targets are not well defined. Should species abundances be conserved at pre-harvest levels of abundance, suggesting that logging should change nothing in the forest? Or should the focus be on maintaining species composition or ecological processes and functions? The chosen target will determine the types of measures that companies are required to implement on the ground to gain certification. Even after years of working within BZP, CIB was intractable on some issues related to wildlife conservation. Road placement in relation to wildlife habitat, selection of set-asides and the size of buffer zones around critical habitat and protected areas were all areas where WCS thought conservation was not given enough consideration. WCS expected the certification process to be an opportunity to make headway on these issues, but the FSC certification process ultimately failed to infuse sufficient rigor into wildlife management considerations in CIB's logging operations. Another shortcoming of certification principles is that they typically focus on endangered species and critical habitat. The protection of endangered species is not a sufficient goal for biodiversity conservation and resource management, particularly where local communities rely on bushmeat as a critical source of protein and income (Poulsen *et al.*, 2009; Poulsen & Clark, 2010).

Lessons learned: Challenges to overcome

Some of the major obstacles to wildlife management and conservation in tropical forests which future PSPCs will likely encounter, and for which solutions need to be found and tested, include the following.

Adding a development partner to the mix

The weakest part of the BZP's multi-faceted approach was the alternative activities program, which failed to produce substantial revenue or protein for local people not employed by the logging company (Poulsen *et al.*, 2007). There are several reasons why the program did not live up to expectations, most important of which was the fact that WCS, CIB and MEF had different goals for the program (see Chapter 4). Even if the BZP partners had had similar expectations however, the program would still have been impeded by technical complexities; there are very few, if any, successful alternative activities projects in Central Africa. The tropical environment renders animal husbandry and agriculture difficult and people are often accustomed to living off of forest resources.

The lesson learned is that improving the livelihoods of rural communities will sometimes require extraordinary measures, which will likely be unique according to differing contexts.

In the CIB concessions, the most straightforward way to improve livelihoods may be to lower barriers for local people to participate in the economy. This could be achieved through preferential hiring of locals and local sourcing of services and goods. If necessary, the company could provide job training to build capacity of local people in positions that require specific technical or administrative skills.

Another way to improve livelihoods may be the incorporation of a development organization with an expertise in alternative livelihood activities into the PSPC. None of the partners (CIB, MEF or WCS) was an expert in sustainable development, and they all had other tasks on which they had to focus. A development partner would add experience, technical know-how and a singular focus to the problem of improving revenue- and protein-producing activities. If industry or government could finance the activities, a local NGO or government agency with development experience might have the advantage of understanding the local environmental and cultural context. In the case of BZP and many other PSPCs, an international development organization that can tap expertise regionally or globally and that can secure additional donor funding might be more effective.

Requiring companies to provide adequate food and meat for their employees and families

The employment of non-local workers can significantly increase the human population in logging concessions and the hunting pressure on wildlife. This is likely to be the situation for other industries where skilled technicians cannot be found in rural areas, especially frontier forests removed from the education opportunities of urban areas. When companies fail to make meat available to employees and their families, these people are forced to rely on bushmeat or freshwater fish. This has two consequences: (1) salaried employees, who are often immigrants, reduce the natural resources on which local, rural people often rely; and (2) the logging company profits twice from the lease of a forest concession – exploitation of both timber and wildlife.

The lesson learned is that companies should have a plan to feed their employees before industrial activities begin, and it should not entail the exploitation of forest resources including wildlife.

If alternative sources of meat cannot be produced locally in an environmentally sustainable way, logging companies should be required to import sufficient meat for employees and their families. To achieve a reduction in the hunting pressure on wildlife, imported products need to be priced equal to or lower than bushmeat so that they represent a competitive alternative.

Reducing the negative impacts of industry and conservation on indigenous people

In addition to prioritizing the rights of indigenous people to land and resources, several other measures should be considered to ensure that indigenous people benefit from industry and conservation. First and foremost, local residents should be given priority for employment by extractive industries. Where local people lack the necessary technical skills, training programs should rectify shortcomings. Investment in capacity may be cost-effective in the long run. Hiring locally would reduce the level of immigration for employment, which should keep the pressure on natural resources closer to pre-industry levels and avoid potential social conflict between indigenous people and immigrants. By keeping the number of immigrants and human population relatively low, the required investment in infrastructure and social services by the company would presumably be lower.

The lesson learned is that companies should have a long-term social plan that sets benchmarks for the employment of local people and limits overall immigration. Whenever possible, the creation of permanent towns should be avoided in timber concessions.

A company's social plan should include steps for mitigating the potential environmental impact of population growth. Immigrants should be encouraged to bring only their immediate family to concessions; additional family members add pressure on social services (sanitation, electricity, medical facilities, etc.) and on natural resources. When immigrant employees conclude their contract, they should be repatriated to their original locale. (CIB provided transport home to all of its employees.) Much of the pressure on natural resources could be avoided by transporting workers into concessions rather than lodging them in camps or towns in the forest. Sawmills and wood-finishing factories should be built and operated in or close to existing cities to avoid growth of urban centers in forests.

When extraction of resources by industry reduces access of local residents to land and resources, they should be compensated. At a minimum, local residents should be guaranteed resource-use areas where the community has sole access rights to fulfill their livelihood needs. In northern Congo, community-use zones are set aside around villages and local residents can use them for agriculture and other forms of resource exploitation. In addition, for every cubic meter of timber cut, CIB puts 300 F CFA (about $0.60) into a community development fund. Each community independently determines how to use its share of the funds. The freedom of the community to choose how it is compensated is the key to any compensation mechanism.

Developing a sustainable long-term funding plan

The BZP has been used as a case study for reproducing the PSPC model, but one aspect that is not replicable is its financial structure. On average, WCS and its donors pay for three-quarters of the annual budget, including contributions to the eco-guard unit and salaries of government officials. This level of financial support by an NGO is unsustainable in the long term. WCS has footed the lion's share of the bill to experiment in the development of a new conservation model and because the NNNP is extremely valuable for conservation. This is unlikely to be the case elsewhere.

The lesson learned is that PSPCs should have short-, medium- and long-term plans to ensure that conservation and management activities are sustainable over time. The vision for BZP was to address the poaching crisis, establish a wildlife management system and, eventually, transfer the majority of the cost burden to the private sector. With the recent global recession, it seems unlikely that this transfer will occur anytime soon (see below).

In theory, the PSPC model should work because industry needs to mitigate its impact on the environment and is willing to pay for assistance in the task. The expense of biodiversity conservation is therefore either offset by more efficient, cost-effective operations or is passed off to customers through higher prices. In some rare cases, the private sector may accept the loss of some part of its profits to 'do the right thing'. PSPCs do open access to financial resources for the private sector and some costs can be covered by partner organizations, but the private sector should be prepared to pay for most of the data collection and management efforts.

Preparing for booms and busts

Economies go through cycles of booms and busts. Thus even though the private sector needs to take on the majority of financing, linking conservation too strongly to the private-sector partner could be a risk. Investment in conservation might be hard to justify to shareholders during difficult financial times, yet stopping the support could set back years of conservation gains. The financial crisis of 2008–2009 resulted in a fall in demand for wood and wood products, particularly in the European market. Logging companies throughout the Congo Basin felt the sharp pinch of the recession. During the crisis, log exporters relied on Asian buyers for stability while overall logging trade was flat with little interest in new purchases. To adapt, companies froze production, shut down sawmills and lived on wood stocks. In northern Congo, CIB shut down all industrial activities in Kabo, stopped one of its sawmills in Pokola and reduced its activities leading to the layoff of 672 national and 7 expatriate workers.

CIB has maintained its commitments to the BZP, a testimony to the program's success and the strength of the partnership. In February 2010 however, the Dalhoff Larsen and Horneman group (DLH) sold CIB to Olam International Limited, leaving the future of wildlife conservation in the area in question. In a news release on 24 February 2011, Olam stated: "We believe a large and growing contiguous FSC-certified natural tropical rain forest, backed by our strong infrastructure and processing capacities in Congo, are the critical foundational enablers for us to be a leading provider of tropical hardwoods globally". Olam therefore seems to be committed to the standards of FSC certification; only time will tell if this commitment includes maintaining the same level of biodiversity conservation and protection of the NNNP.

The lesson learned is that financial contingency plans need to be developed for PSPC's. One solution is for partnerships to invest in a trust fund during good economic times that could keep it going through rough times. The NGO could raise funds from its donors or seek funds through intergovernmental donors. The company could dedicate a small percentage of its revenues to the trust fund. The government could rebate part of the fees charged the concessionaire. Perhaps one way to attract international donors is promise that the company and government will match donations up to some maximum contribution. A trust fund might be a particularly good investment in industries that work on long time horizons and have long-term leases to land.

Setting region-wide standards through enforcement of national laws

There is a trend towards better land-use planning and forest management in Central Africa. Central African governments have recognized the need for management plans for concessions and, at least in the case of Congo, the existing forestry laws correspond to or even surpass internationally recognized standards. Moreover, the Congolese government is slowly starting to enforce its own legislation: nine management plans are advanced in their development, including the Kabo, Pokola, and Loundoungou concessions that have been adopted and received FSC certification. Of the 69 forest management units in Republic of Congo, 50% are committed to the process of sustainable forest management planning (Box 7.2).

Box 7.2 Expansion of Forestry and Wildlife management across Central Africa (John R. Poulsen, Duke University)

The idea of an NGO partnering with a logging company and government to achieve biodiversity conservation would have been preposterous 15 or 20 years ago. Now the Buffer Zone Project (BZP) is one of many such partnerships that work in numerous industries including eco-tourism, mining, fishing and agriculture. The BZP stands apart, however, in several ways. First, the extent of the activities that the partners undertook together is remarkable: they worked together daily to carry out the management of wildlife and forests over an enormous landscape. Second, the partnership has been long lasting, and is entering its 13th year. Third, the BZP has demonstrated discernible outcomes. There is strong evidence that the partnership has succeeded in mitigating the impacts of logging on wildlife. Data demonstrate that the NNNP has been successfully protected from poaching. The certification of the CIB forestry procedures and products by the FSC is proof of improved logging techniques. Finally, there is a strong knowledge of why the partnership has been successful (Elkan & Elkan, 2005; Elkan *et al.*, 2006; Poulsen *et al.*, 2007, 2009; Poulsen & Clark, 2010).

Because BZP is unique in all these aspects, the private sector and conservation community should capitalize on BZP's investment and experience to organize it as a *center of excellence for training in PSPCs* with a view to replicate the model through the Republic of Congo, the Congo Basin and beyond. The BZP could train foresters, conservationists and managers, largely by integrating them into the activities in which it is already engaged. The knowledge and experiences could then be transferred and adapted to expand responsible forestry and wildlife management throughout the tropics.

The lesson learned is that PSPCs will only work in a national context of improved and enforced regulations. Otherwise, some companies will act as cheaters, gaining greater profits by foregoing the costs of being responsible. Cheating leads to leakage where the environmental gains in an area under good management are lost through displacement to an area with no management because the bad actor shirks its responsibilities.

Certain companies such as CIB have made considerable investments in infrastructure and procedures to promote sustainable forest management, social development and wildlife management. To promote land management and conservation at a regional scale however, forestry laws should be applied to all companies and all concessions without exception – Central African countries need to enforce their own laws.

Conclusion

This book describes a new model of conservation born from the threat of logging to tropical forest, its biodiversity and the livelihoods of its peoples. At the time of its conception, the BZP was novel and radical; most importantly, it was a practical solution to a conservation problem. Results presented here suggest that the model is effective and its replication to other sites and other industries is likely to help conserve tropical habitats, their biodiversity and their wildlife. After years of work and experimentation, many important lessons can be drawn from BZP. One lesson rings truer than any other, however: biodiversity conservation in a changing world requires innovative thinking and audacious efforts. As the reach of industry expands in the coming decades, the most important conservation might be conducted outside protected areas and will likely include partnerships with the private sector that were once thought improbable.

References

Agrawal, A. and Ostrom E. (2001) Collective action, property rights and decentralization in resource use in India and Nepal. *Politics and Society*, **29**, 485–514.

Albrechtsen, L., David, W.M.A., Paul, J.J., Ramon, C. and Fa, J.E. (2007) Faunal loss from bushmeat hunting, empirical evidence and policy implications in Bioko island. *Environmental Science & Policy*, **10**, 654–667.

Armitage, D. and Johnson, D. (2006) Can resilience be reconciled with globalization and the increasingly complex conditions of resource degradation in Asian coastal regions? *Ecology and Society*, **11**, http://www.ecologyandsociety.org/vol11/iss1/art2/.

Arnhem, E., Dupain, J., Vercauteren Drubbel, R., Devos, C. and Vercauteren, M. (2008) Selective logging, habitat quality and home range use by sympatric gorillas and chimpanzees. A case study from an active logging concession in southeast Cameroon. *Folia Primatologica*, **79**, 1–14.

Asner, G.P., Broadbent, E.N., Oliveira, P.J.C., Keller, M., Knapp, D.E. and Silva, J.N.M. (2006) Condition and fate of logged forests in the Brazilian Amazon. *Proceedings of National Academy of Sciences, USA*, **103**, 12947–12950.

Auzel, P. and Wilkie, D.S. (2000) Wildlife use in northern Congo: Hunting in a commercial logging concession, in *Hunting for Sustainability in Tropical Forests* (eds J.G. Robinson and E.L. Bennett). Columbia University Press, New York, 413–426.

Azevedo-Ramos, C., de Carvalho, O. and do Amaral, B.D. (2006) Short-term effects of reduced-impact logging on eastern Amazon fauna. *Forest Ecology & Management*, **232**, 26–35.

Barnes, R.F.W. (2001) How reliable are dung counts for estimating elephant numbers? *African Journal of Ecology*, **39**, 1–9.

Barreto, P., Souza, C., Nogueron, R., Anderson, A. and Salomao, R. (2006) *Human Pressure on the Brazilian Amazon Forests*. World Resources Institute, Washington, DC.

Barrett, C.B., Brandon, K., Gibson, C. and Gjertsen, H. (2001) Conserving tropical biodiversity amid weak institutions. *Bioscience*, **51**, 497–502.

Tropical Forest Conservation and Industry Partnership: An Experience from the Congo Basin, First Edition. Edited by Connie J. Clark and John R. Poulsen.
© 2012 Wildlife Conservation Society. Published 2012 by John Wiley & Sons, Ltd.

Basset, Y., Charles, E., Hammond, D.S. and Brown, V.K. (2001) Short-term effects of canopy openness on insect herbivores in a rain forest in Guyana. *Journal of Applied Ecology*, **38**, 1045–1058.

Bawa, K.S. and Seidler, R. (1998) Natural forest management and conservation of biodiversity in tropical forests. *Conservation Biology*, **12**, 46–55.

Bennett, E.L. (2001) Timber certification: Where is the voice of the biologist? *Conservation Biology*, **15**, 308–310.

Bennett, E.L. Blencowe, K., Brown, D., Burn, R.W., Cowlishaw, G., Davies, G., Dublin, H., Fa, J.E., Milner-Gulland, E.J., Robinson, J.G., Rowcliffe, J.M., Underwood, F.M. and Wilkie, D.S. (2007) Hunting for consensus: Reconciling bushmeat harvest, conservation, and development policy in west and central Africa. *Conservation Biology*, **21**, 884–887.

Berkes, F. (2006) From community-based resource management to complex systems: The scale issue and marine commons. *Ecology and Society*, **11**.

Bhagwat, S.A., Willis, K.J., Birks, H.J.B. and Whittaker, R.J. (2008) Agroforestry, a refuge for tropical biodiversity? *Trends in Ecology & Evolution*, **23**, 261–267.

Blake, S. (2002) Forest buffalo prefer clearings to closed-canopy forest in the primary forest of northern Congo. *Oryx*, **36**, 81–86.

Blake, S. and Inkamba-Nkulu, C. (2004) Fruit, minerals, and forest elephant trails: Do all roads lead to Rome? *Biotropica*, **36**, 392–401.

Blake, S., Strindberg, S., Boudjan, P., Makombo, C., Bila-Isia, I., Ilambu, O., Grossmann, F., Bene-Bene, L., de Semboli, B., Mbenzo, V., S'hwa, D., Bayogo, R., Williamson, L., Fay, M., Hart, J. and Maisels, F. (2007) Forest elephant crisis in the Congo Basin. *Plos Biology*, **5**, 945–953.

Blake, S., Deem, S.L., Strindberg, S., Maisels, F., Momont, L., Bila-Isia, I., Douglas-Hamilton, I., Karesh, W.B. and Kock, M.D. (2008) Roadless wilderness area determines forest elephant movements in the Congo Basin. *PLoS ONE*, **3**, e3546.

Boen-Jones, E., Brown, D. and Robinson, E.J.Z. (2003) Economic commodity or environmental crisis? An interdisciplinary approach to analysing the bushmeat trade in central and west Africa. *Area*, **35**, 390–402.

Borrini-Feyerabend, G. (1996) *Collaborative Management of Protected Areas, Tailoring the Approach to the Context*. IUCN, Gland, Switzerland.

Bowland, A.E. and Perrin, M.R. (1995) Temporal and spatial patterns in blue duikers Philantomba monticola and red duikers Cephalophus natalensis. *Journal of Zoology, London*, **237**, 487–498.

Brashares, J.S., Arcese, P., Sam, M.K., Coppolillo, P.B., Sinclair, A.R.E. and Balmford, A. (2004) Bushmeat hunting, wildlife declines, and fish supply in West Africa. *Science*, **306**, 1180–1183.

Brncic, T.M., Willis, K.J., Harris, D.J., Telfer, M.W. and Bailey, R.M. (2009) Fire and climate change impacts on lowland forest composition in northern Congo during the last 2580 years from palaeoecological analyses of a seasonally flooded swamp. *Holocene*, **19**, 79–89.

Brown, D. (2007) Is the best the enemy of the good? Institutional and livelihoods perspectives on bushmeat harvesting and trade – some issues and challenges, in *Bushmeat and Livelihoods, Wildlife Management and Poverty Reduction* (eds G. Davies and D. Brown). Blackwell Publishing Ltd, Oxford, 111–124.

Buckland, S.T., Anderson, D.R., Burnham, K.P., Laake, J.L., Borchers, D.L. and Thomas, L. (2001) *Introduction to Distance Sampling: Estimating Abundance of Biological Populations*. Oxford University Press, Oxford, UK.

Butler, R.A. and Laurance, W.F. (2008) New strategies for conserving tropical forests. *Trends in Ecology & Evolution*, **23**, 469–472.

Carlsson, L. and Berkes, F. (2005) Co-management, concepts and methodological implications. *Journal of Environmental Management*, **75**, 65–76.

Castillo, O., Clark, C.J., Coppolillo, P., Kretser, H., McNab, R., Noss, A., Quieroz, H., Tessema, Y., Veddar, A., Wallace, R., Walston, J. and Wilkie, D. (2006) Casting for conservation actors, people, partnerships and wildlife, in *WCS Working Paper*. WCS, New York.

CBFP (Congo Basin Forest Partnership) (2006) *The Forests of the Congo Basin, State of the Forest 2006*. CBFP, Bonn, Germany.

Chapin, F.S., Zavaleta, E.S., Eviner, V.T., Naylor, R.L., Vitousek, P.M., Reynolds, H.L., Hooper, D.U., Lavorel, S., Sala, O.E., Hobbie, S.E., Mack, M.C. and Diaz, S. (2000) Consequences of changing biodiversity. *Nature*, **405**, 234–242.

Chapman, C.A., Struhsaker, T.T. and Lambert, J.E. (2005) Thirty years of research in Kibale National Park, Uganda, reveals a complex picture for conservation. *International Journal of Primatology*, **26**, 539–555.

Chazdon, R.L. (1998) Ecology: Tropical forests: Log 'em or leave 'em? *Science*, **281**, 1295–1296.

Chazdon, R.L. Peres, C.A., Dent, D., Sheil, D., Lugo, A.E., Lamb, D., Stork, N.E., Miller, S.E. (2009) The potential for species conservation in tropical secondary forests. *Conservation Biology*, **23**, 1406–1417.

Clark, C.J. and Elkan, S. (2004) *Connaissez-vous les Grands Mammifères Protégés de la République du Congo?* WCS, New York.

Clark, C.J., Poulsen, J.R., Malonga, R. and Elkan, P.W. (2009) Logging concessions can extend the conservation estate for central African tropical forests. *Conservation Biology*, **23**, 1281–1293.

Cleary, D.F.R., Boyle, T.J.B., Setyawati, T., Anggraeni, C.D., Van Loon, E.E., Menken, S.B.J. (2007) Bird species and traits associated with logged and unlogged forest in Borneo. *Ecological Applications*, **17**, 1184–1197.

Congolaise Industrielle des Bois (2006) Plan d'amenagement de l'unité forestière d'aménagement de Kabo (2005–2034). Republic of Congo, Ministry of Forest Economy.

Congolaise Industrielle des Bois (2008) Plan d'amenagement de l'unité forestière d'aménagement de Pokola. Republic of Congo, Ministry of Forest Economy.

Corlett, R.T. (2007) The impact of hunting on the mammalian fauna of tropical Asian forests. *Biotropica*, **39**, 292–303.

Cowlishaw, G., Mendelson, S. and Rowcliffe, J.M. (2005) Evidence for post-depletion sustainability in a mature bushmeat market. *Journal of Applied Ecology*, **42**, 460–468.

Cowlishaw, G., Mendelson, S. and Rowcliffe, J.M. (2007) Livelihoods and sustainability in a bushmeat commodity chain in Ghana, in *Bushmeat and Livelihoods, Wildlife Management and Poverty Reduction* (eds G. Davies and D. Brown). Blackwell Publishing, Oxford, 32–46.

Craig, M.D. and Roberts, J.D. (2005) The short-term impacts of logging on the jarrah forest avifauna in southwest Western Australia, implications for the design and analysis of logging experiments. *Biological Conservation*, **124**, 177–188.

Crome, F.H.J., Thomas, M.R. and Moore, L.A. (1996) A novel Bayesian approach to assessing impacts of rain forest logging. *Ecological Applications*, **6**, 1104–1123.

Crookes, D.J., Ankudey, N. and Milner-Gulland, E.J. (2005) The value of a long-term bushmeat market dataset as an indicator of system dynamics. *Environmental Conservation*, **32**, 333–339.

Curran, B., Sunderland, T., Maisels, F., Oates, J., Asaha, S., Balinga, M., Defo, L., Dunn, A., Telfer, P., Usongo, L., von Loebenstein, K. and Roth, P. (2009) Are central Africa's protected areas displacing hundreds of thousands of rural poor? *Conservation and Society*, **7**, 30–45.

Cyranoski, D. (2007) Logging, the new conservation. *Nature*, **446**, 608–610.

Davies, G. (2002) Bushmeat and international development. *Conservation Biology*, **16**, 587–589.

de Merode, E., Homewood, K. and Cowlishaw, G. (2004) The value of bushmeat and other wild foods to rural households living in extreme poverty in Democratic Republic of Congo. *Biological Conservation*, **118**, 573–581.

de Queiroz, J., App, B.B., Morin, R. and Rice, W. (2008) Partnering with extractive industries for the conservation of biodiversity in Africa, A guide for USAID engagement. Biodiversity Assessment and Technical Support Program (BATS) EPIQ IQC: EPP-I-00-03-00014-00, Task Order 02.

de Souza-Mazurek, R.R., Pedrinho, T., Feliciano, X., Hilário, W., Gerôncio, S. and Marcelo, E. (2000) Subsistence hunting among the Waimiri Atroari Indians in central Amazonia, Brazil. *Biodiversity & Conservation*, **9**, 579–596.

Dubost, G. (1980) L'écologie et la vie sociale du Céphalophe bleu (Cephalophus monticola Thunberg), petit ruminant forestrier atricain. *Zeltschrifl Fur Tierpsychologie*, **54**, 205–266.

Dunn, R.R. (2004) Managing the tropical landscape, a comparison of the effects of logging and forest conversion to agriculture on ants, birds, and lepidoptera. *Forest Ecology & Management*, **191**, 215–224.

Eaton, M.J. (2002) Subsistence wildlife hunting in a multi-use forest of the Republic of Congo, monitoring and management for sustainable harvest Minneapolis. MSc thesis, University of Minnesota.

Eaton, MJ. (2006) Assessment of crocodile populations in Mayumba National Park and associated waterways, Republic of Gabon. Report to the Wildlife Conservation Society, Bronx, NY.

Eaton, M.J., Thorbjarnarson, A.P., Amato, G. (2009) Species-level diversification of African dwarf crocodiles (Genus Osteolaemus), a geographic and phylogenetic perspective. *Molecular Phylogenetics and Evolution*, **50**, 496–506.

Eba'a Atyi, R., Devers, D., de Wasseige, C. and Maisels, F. (2009) State of the forests of Central Africa, regional synthesis, in *The Forests of the Congo Basin – State of the Forest 2008* (eds C. de Wasseige, D. Devers, P. de Marcken, R. Eba'a Atyi, R. Nasi and P. Mayaux). Congo Basin Forest Partnership, Luxembourg, 15–41.

Edmunds, D. and Wollenberg, E. (2002) Disadvantaged groups in multistakeholder negotiations. CIFOR program report, Bogor, Indonesia.

Efoakondza, B. (1993) Mensurations, comptage, pesage et commercialization du crocodile nain dans le nord du pays (Congo) Osteolaemus tetraspis. WCS, Brazzaville, Congo.

Elkan, P.W. (2003) Ecology and conservation of bongo antelope (*Tragelaphus euryceros*) in lowland forest, northern Republic of Congo (2003). PhD dissertation, University of Minnesota.

Elkan, P. and Elkan, S. (2005) Mainstreaming wildlife conservation in multiple-use forests of northern Republic of Congo, in *Mainstreaming Conservation in Multiple-Use Landscapes* (eds C. Peterson and B. Huntley). GEFSEC Publications, Washington, DC, 51–67.

Elkan, P.W., Elkan, S.W., Moukassa, A., Malonga, R., Ngangoué, M. and Smith, J.L.D. (2006) Managing threats from bushmeat hunting in a timber concession in the Republic of Congo, in *Emerging Threats to Tropical Forests* (eds C. Peres and W. Laurence) University of Chicago Press, Chicago, 395–415.

Emery-Thompson, M., Kahlenberg, S., Gilby, I. and Wrangham, R.W. (2007) Core area quality is associated with variance in reproductive success among female chimpanzees at Kanyawara, Kibale, National Park. *Animal Behaviour*, **73**, 501–512.

Eves, H.E. and Ruggiero, R.G. (2000) Socio-economics and the sustainability of hunting in the forests of northern Congo (Brazzaville), in *Hunting for Sustainability in Tropical Forests* (eds J.G. Robinson and E.L. Bennett). Columbia University Press, New York, 427–454.

Fa, J.E. and Peres, C.A. (2001) Game vertebrate extraction in African and neotropical forests, an intercontinental comparison, in *Conservation of Exploited Species* (eds J.D. Reynolds, G.M. Mace, J.G. Robinson and K.H. Redford). Cambridge University Press, Cambridge, 203–224.

Fa, J.E., Yuste, J.E.G. and Castelo, R. (2000) Bushmeat markets on Bioko Island as a measure of hunting pressure. *Conservation Biology*, **14**, 1602–1613.

Fa, J.E., Peres, C.A. and Meeuwig, J. (2002) Bushmeat exploitation in tropical forests, an intercontinental comparison. *Conservation Biology*, **16**, 232–237.

Fa, J.E., Ryan, S.F. and Bell, D.J. (2005) Hunting vulnerability, ecological characteristics and harvest rates of bushmeat species in afrotropical forests. *Biological Conservation*, **121**, 167–176.

Fagan, C.C., Peres, A. and Terborgh, J. (2006) Tropical forests, a protected-area strategy for the twenty-first century. In *Emerging Threats to Tropical Forests* (eds Peres, C.A. and Laurance, W.F.), University of Chicago Press, Chicago, 417–434.

FAO (2009) State of the World's Forests. FAO, Rome.

Fearnside, P.M. (2001) Soybean cultivation as a threat to the environment in Brazil. *Environmental Conservation*, **28**, 23–38.

Ferraro, P.J. and Kramer, R.A. (1995) A framework for affecting household behavior to promote biodiversity conservation. Environment and Natural Resource Policy and Training (EPAT) Project. Winrock International Environmental Alliance, Arlington, VA.

Florin, P. and Wandersman, A. (1990) Citizen participation, voluntary organizations and community development: Insights for empowerment through research. *American Journal of Community Psychology*, **18**, 41–55.

FSC (2006) FSC principles and criteria for forest stewardship FSC, Bonn, Germany.

Gami, N. and Doumenge, C. (2001) Les acteurs de la gestion forestière en Afrique centrale et de l'Ouest. Forafri, Libreville, Gabon.

Ghimire, K.B. and Pimbert, M.P. (1997) *Social Change and Conservation*. Earthscan Publications Limited, London.

Gibson, C.C., McKean, M.A. and Ostrom, E. (2000) *People and Forests, Communities, Institutions and Governance*. MIT Press, Cambridge, MA.

Global Forest Watch (2002) An analysis of access into Central Africa's rainforests. World Resources Institute, Washington, DC.

Grubb, P. (2002) Review of Viv Wilson's Duikers of Africa (Masters of the African Forest Floor). *Gnusletter*, **21**: 2–3. Accessed October 2010 (http://cmsdata.iucn.org/downloads/vol_21_n2_2002_all.pdf).

Gullison, R.E., Frumhoff, P.C., Canadell, J.G., Field, C.B., Nepstad, D.C., Hayhoe, K., Avissar, R., Curran, L.M., Friedlingstein, P., Jones, C.D. and Nobre, C. 2007. Tropical forests and climate policy. *Science*, **316**, 985–986.

Hackel, J.D. (1999) Community conservation and the future of Africa's wildlife. *Conservation Biology*, **13**, 726–734.

Hansen, M.C., Stehman, S.V. and Potapova, P.V. (2010) Quantification of global gross forest cover loss. *Proceedings of National Academy of Sciences, USA*, **107**, 8650–8655.

Hardin, R. (2011) Concessionary Politics: Property, Patronage and Political Rivalry in Central African Forest Management. *Current Anthropology*, **52**, S113–S125.

Harris, D.J. (2002) *The Vascular Plants of the Dzanga–Sangha Reserve, Central African Republic*. National Botanic Garden, Meisse, Belgium.

Harris, D.J. and Wortley, A.H. (2008) *Sangha Trees, an Illustrated Identification Manual*. Royal Botanic Gardens, Edinburgh, UK.

Hashimoto, C. (1995) Population census of the chimpanzees in the Kalinzu Forest, Uganda: Comparison between methods with nest counts. *Primates*, **36**, 477–488.

Hekkala, E., Shirley, M.H., Amato, G., Austin, J.D., Charter, S., Thorbjarnarson, J., Vliet, K.A., Houck, M.L., Desalle, R. and Blum, M.J. (2011) An ancient icon reveals new mysteries: Mummy DNA resurrects a cryptic species within the Nile crocodile. *Molecular Ecology*, **20**, 4199–4215.

Heydon, M.J. and Bulloh, P. (1997) Mousedeer densities in a tropical rainforest: The impact of selective logging. *Journal of Applied Ecology*, **34**, 484–496.

Holdsworth, A.R. and Uhl, C. (1997) Fire in Amazonian selectively logged rain forest and the potential for fire reduction. *Ecological Applications*, **7**, 713–725.

Holt, F.L. (2005) The Catch-22 of conservation, indigenous peoples, biologists, and cultural change. *Human Ecology*, **33**, 199–215.

Hooge, P.N. and Eichenlaub, B. (2000) Animal movement extension to ArcView. Alaska Science Center, Biological Science Office, US Geological Survey, Anchorage, AK, USA (USGS–BRD).

Houghton, R.A. (2008) Biomass, in *Encyclopedia of Ecology*, 1st edition (eds S.E. Jorgensen and B.D. Fath), Oxford, Elsevier, 448–453.

Hurst, A. (2007) Institutional challenges to sustainable bushmeat management in central Africa, in *Bushmeat and Livelihoods, Wildlife Management and Poverty Reduction* (eds G. Davies and D. Brown), Blackwell Publishing, Oxford, 158–172.

Hutton, J., Adams, W.M. and Murombedzi, J.C. (2005) Back to the barriers? Changing narratives in biodiversity conservation. *Forum for Development Studies*, **32**, 341–370.

Intergovernmental Panel on Climate Change (2007) The Physical Science Basis. Contribution of working group I to the fourth assessment report of the IPCC. Cambridge University Press, Cambridge, UK.

ITTO (2005) Revised ITTO criteria and indicators for the sustainable management of tropical forests, in *ITTO Policy Development Series*, ITTO, Yokohama.

IUCN (1996) Resolution 1.42 on Collaborative Management for Conservation. IUCN, Gland, Switzerland.

IUCN (2006) Guidelines for the conservation and sustainable use of biodiversity in tropical timber production forests. IUCN, Gland, Switzerland.

Jerozolimski, A. and Peres, C.A. (2003) Bringing home the biggest bacon, a cross-site analysis of the structure of hunter-kill profiles in Neotropical forests. *Biological Conservation*, **111**, 415–425.

Johns, J.S., Barreto, P. and Uhl, C. (1996) Logging damage during planned and unplanned logging operations in the eastern Amazon. *Forest Ecology & Management*, **89**, 59–77.

Kaimowitz, D., Thiele, G. and Pacheco, P. (1999) The effects of structural adjustment on deforestation and forest degradation in lowland Bolivia. *World Development*, **27**, 505–520.

Kernohan, B.J., Gitzen, R.A. and Millspaugh, J.J. (2001) Analysis of animal space use and movements, in *Radio Tracking and Animal Populations* (eds J.J. Millspaugh and J.M. Marzluff). Academic Press, San Diego, California, 125–166.

Kingdon, J. (1997) *The Kingdon Field Guide to African Mammals*. Academic Press, London.

Kirilenko, A.P. and Sedjo, R.A. (2007) Climate change impacts on forestry. *Proceedings of National Academy of Sciences, USA*, **104**, 19697–19702.

Kitanishi, K. (1995) Seasonal changes in the subsistence activities and food intake of the aka hunter–gatherers in northeastern Congo. *African Study Monographs*, **16**, 73–118.

Kooiman, J. (2003) *Governing as Governance*. SAGE Publications Ltd, London.

Kreulen, D.A. (1985) Lick use by large herbivores, a review of benefits and banes of soil consumption. *Mammal Review*, **15**, 107–123.

Lamb, D, Erskine, P.D. and Parrotta, J.A. (2005) Restoration of degraded tropical forest landscapes. *Science*, **310**, 1628–1632.

Lanfranchi, R., Ndanga, J. and Zana, H. (1998) New carbon 14C datings of iron metallurgy in the Central African dense forest, in *Resource Use in the Trinational Sangha River Region of Equatorial Africa: Histories, Knowledge Forms, and Institutions* (ed. H.E. Eves). School of Forestry & Environmental Studies, New Haven, CT, Yale, 41–50.

Laporte, N.T., Stabach, J.A., Grosch, R., Lin, T.S. and Goetz, S.J. (2007) Expansion of industrial logging in Central Africa. *Science*, **316**, 1451–1451.

Laurance, W.F. (2000) Cut and run, the dramatic rise of transnational logging in the tropics. *Trends in Ecology & Evolution*, **15**, 433–434.

Laurance, W.F. and Peres, C.A. (2006) *Emerging Threats to Tropical Forests*. University of Chicago Press, Chicago and London.

Laurance, W.F., Croes, B.M., Tchignoumba, L., Lahm, S.A., Alonso, A., Lee, M.E., Campbell, P. and Ondzeano, C. (2006) Impacts of roads and hunting on central African rainforest mammals. *Conservation Biology*, **20**, 1251–1261.

Lewis, J. (2002) Forest hunter-gatherers and their world, a study of the Mbendjele Yaka Pygmies of Congo-Brazzaville and their secular and religious activities and representations. London School of Economics and Political Science, University of London.

Malanda, G.A.F., Iyenguet, F.C., Ntalassani, F., Madzoké, B.S.D. (2005) Étude des grands mammifères dans l'ouest de la Bailly. WCS, Republic of Congo.

Malcolm, J.R. and Ray, J.C. (2000) Influence of timber extraction routes on central African small mammal communities, forest structure, and tree diversity. *Conservation Biology*, **14**, 1623–1638.

Malhi, Y. and Grace, J. (2000) Tropical forests and atmospheric carbon dioxide. *Trends in Ecology and Evolution*, **15**, 332–337.

Martin, T.G., Wintle, B.A., Rhodes, J.R., Kuhnert, P.M., Field, S.A., Low-Choy, S.J., Tyre, A.J. and Possingham, H.P. (2005) Zero tolerance ecology: improving ecological inference by modeling the source of zero observations. *Ecology Letters*, **8**, 1235–1246.

Matthews, A. and Matthews, A. (2004) Survey of gorillas (Gorilla gorilla gorilla) and chimpanzees (Pan troglodytes troglodytes) in Southwestern Cameroon. *Primates*, **45**, 15–24.

Mavah, G. (2005) Synthese demographique des villages et campements dans et en peripherie des UFA de Pokola, Loundoungou et de Toukoulaka Kabo, WCS, Republic of Congo.

Mayers, J. and Vermeulen, S. (2002) Company-community forestry partnerships: From raw deals to mutual gains? In *Instruments for Sustainable Private Sector Forestry Series*. International Institute for Environment and Development (IIED), London.

Meijaard, E., Sheil, D., Nasi, R. and Stanley, S.A. (2006) Wildlife conservation in Bornean timber concessions. *Ecology and Society*, **11**, 47.

Milner-Gulland, E.J., Bennett, E.L. and the SCB Annual Meeting Wild Meat Group. (2003) Wild meat, the bigger picture. *Trends in Ecology & Evolution*, **18**, 351–357.

Mockrin, M.H. (2008) *The Spatial Structure and Sustainability of Subsistence Wildlife Harvesting in Kabo, Congo Columbia*. PhD dissertation, Columbia University, New York.

Morgan, D. and Sanz, C. (2003) Naive encounters with chimpanzees in the Goualougo Triangle, Republic of Congo. *International Journal of Primatology*, **24**, 369–381.

Morgan, D., Sanz, C., Onononga, J.R. and Strindberg, S. (2006) Ape abundance and habitat use in the Goualougo Triangle, Republic of Congo. *International Journal of Primatology*, **27**, 147–179.

Moukassa, A. (2004) Se nourrir dans un camp forestier, suivi de l'alimentation des menages dans les sites forestiers de Kabo et de Ndoki 2 Kabo. PROGEPP, Republic of Congo.

Moukassa, A. and Kimbembe, B. (2003) Utilisation de l'espace forestier par les communautes des terres Mouzoumou Kabo. PROGEPP, Republic of Congo.

Moukassa, A., Nsosso, D. and Mavah, G. (2005) Occupation de l'espace forestier par les communautes villageoises et semi–nomades dans les UFA Kabo, Pokola, Toukoulaka et Loundoungou (Foret Nord–Congo) Kabo. PROGEPP, Republic of Congo.

Naughton-Treves, L. and Weber, W. (2001) Human dimensions of the African rain forest, in *African Rain Forest Ecology and Conservation* (eds W. Weber, L.J.T. White, A. Vedder and N. Naughton-Treves). Yale University Press, New Haven, CT, 22–46.

Nelson. A and Chomitz, K.M. (2009) Protected area effectiveness in reducing tropical deforestation: A global analysis of the impact of protection status. *Independent Evaluation Group Evaluation Brief 7*, The World Bank, Washington, DC, 253–260, http://sitesources.worldbank.org/INTOED/Resources/.

Nepstad, D.C., Verissimo, A., Alencar, A., Nobre, C., Lima, E., Lefebvre, P., Schlesinger, P., Potter, C., Moutinho, P., Mendoza, E., Cochrane, M. and Brooks, V. (1999) Large-scale impoverishment of Amazonian forests by logging and fire. *Nature*, **398**, 505–508.

Nepstad, D.C., Stickler, C.M. and Almeida, O.T. (2006) Globalization of the Amazon soy and beef industries, opportunities for conservation. *Conservation Biology*, **20**, 1595–1603.

Noss, A.J. (1998) The impacts of BaAka net hunting on rainforest wildlife. *Biological Conservation*, **86**, 161–167.

Novaro, A.J., Redford, K.H. and Bodmer, R.E. (2000) Effect of hunting in source-sink systems in the neotropics. *Conservation Biology*, **14**, 713–721.

Parry, L., Barlow, J. and Peres, C.A. (2009) Allocation of hunting effort by Amazonian smallholders, implications for conserving wildlife in mixed-use landscapes. *Biological Conservation*, **142**, 1777–1786.

Pearce, D., Putz, F.E. and Vanclay, J.K. (2003) Sustainable forestry in the tropics, panacea or folly? *Forest Ecology & Management*, **172**, 229–247.

Peres, C.A. and Palacios, E. (2007) Basin-wide effects of game harvest on vertebrate population densities in Amazonian forests, implications for animal-mediated seed dispersal. *Biotropica*, **39**, 304–315.

Peres, C.A., Barlow, J. and Laurance, W.F. 2006. Detecting anthropogenic disturbance in tropical forests. *Trends in Ecology and Evolution*, **21**, 227–229.

Pimm, S.L. and Lawton, J.H. (1998) Ecology: Planning for biodiversity. *Science*, **279**, 2068–2069.

Plumptre, A.J. and Reynolds, V. (1994) The effect of selective logging on the primate populations in the Budongo Forest Reserve. *Uganda Journal of Applied Ecology*, **31**, 631–641.

Poulsen, J.R. (2009) Building private-sector partnerships for conservation (PSPCs): Lessons learned from the collaboration between WCS, CIB, and the Republic of Congo in forestry concessions. USAID, Washington, DC.

Poulsen, J.R. and Clark, C.J. (2004) Densities, distributions, and seasonal movements of gorillas and chimpanzees in swamp forest in northern Congo. *International Journal of Primatology*, **25**, 285–306.

Poulsen, J.R. and Clark, C.J. (2010) Congo Basin timber certification and biodiversity conservation, in *Biodiversity Conservation in Certified Forests* (eds D. Sheil, F.E. Putz and R.J. Zagt). Tropenbos International, Wageningen, Netherlands, 55–60.

Poulsen, J.R., Clark, C.J., Connor, E.F. and Smith, T.B. (2002) Differential resource use by primates and hornbills, implications for seed dispersal. *Ecology*, **83**, 228–240.

Poulsen, J., Clark, C. and Malonga, R. (2004a) Recensement et distribution des grands mammifères dans l'unité forestière d'aménagement de Kabo (Nord Congo). PROGEPP, République du Congo.

Poulsen, J., Clark, C. and Malonga, R. (2004b) Recensement et distribution des grands mammifères dans l'unité forestière d'aménagement de Pokola (Nord Congo). PROGEPP, République du Congo.

Poulsen, J., Clark, C. and Malonga, R. (2004c) Recensement et distribution des grands mammifères dans l'unité forestière d'aménagement de Loundougou (Nord Congo). PROGEPP, République du Congo.

Poulsen, J., Clark, C. and Malonga, R. (2004d) Recensement et distribution des grands mammifères dans l'unité forestière d'aménagement de Toukoulaka (Nord Congo). PROGEPP, République du Congo.

Poulsen, J.R., Clark, C.J. and Mavah, G. (2007) Wildlife management in a logging concession in northern Congo: Can livelihoods be maintained through sustainable hunting? in *Bushmeat and Livelihoods* (eds G. Davies and D. Brown). Blackwell Publishing, Oxford, 140–157.

Poulsen, J.R., Clark, C.J., Mavah, G. and Elkan, P.W. (2009) Bushmeat supply and consumption in a tropical logging concession in northern Congo. *Conservation Biology*, **23**, 1597–1608.

Poulsen, J.R., Clark, C.J. and Bolker, B.M. 2011. Decoupling the effects of logging and hunting on an Afrotropical animal community. *Ecological Applications*, **21**, 1819–1836.

PROGEPP (1999) Protocole d'accord sur la gestion des ecosystèmes périphériques au Parc National de Nouabalé–Ndoki, Nord Congo entre le MEF, CIB, WCS, et Congo Safari, PROGEPP, République du Congo.

PROGEPP (2005a) *Procedures de Zonage pour la Gestion de la Faune*. PROGEPP, République du Congo.

PROGEPP (2005b) *Procedures Socio-economiques dans les UFA de Pokola, Kabo et Loundoungou*. PROGEPP, République du Congo.

PROGEPP (2005c) *Procedures de Suivi de la Faune*. PROGEPP, République du Congo.

PROGEPP (2008) Protocole d'accord relative à la mise en œuvre du Projet de Gestion des Ecosystèmes Périphériques au Parc National de Nouabalé–Ndoki

(PROGEPP) dans les concessions forestières attribuées à la CIB. PROGEPP, République du Congo.

Putz, F.E., Redford, K.H., Fimbel, R., Robinson, J.G. and Blate, G.M. (2000) Biodiversity conservation in the context of tropical forest management, in *Environment Paper*. World Bank, Washington, DC.

Putz, F.E., Blate, G.M., Redford, K.H., Fimbel, R. and Robinson, J. (2001) Tropical forest management and conservation of biodiversity, an overview. *Conservation Biology*, **15**, 7–20.

Putz, F.E., Sist, P., Fredericksen, T., Dykstra, D. (2008) Improved tropical forest management for carbon retention. *Plos Biology*, **6**, e166.

Raunikar, R., Buongiorno, J., Turner, J.A. and Zhu, S. (2009) Global outlook for wood and forests with the bioenergy demand implied by scenarios of the intergovernmental panel on climate change. *Forest Policy & Economics*, **12**: 48–56.

Rice, R.E., Gullison, R.E. and Reid, J.W. (1997) Can sustainable management save tropical forests? *Scientific American*, **276**, 44–49.

Rist, J., Rowcliffe, M., Cowlishaw, G. and Milner-Gulland, E.J. (2008) Evaluating measures of hunting effort in a bushmeat system. *Biological Conservation*, **141**, 2086–2099.

Robinson, J.G., Redford, K.H. and Bennett, E.L. (1999) Conservation: Wildlife harvest in logged tropical forests. *Science*, **284**, 595–596.

Rose, C.M. (2000) Common property, regulatory property, and environmental protection, comparing community-based management to tradable environmental allowances, in *The Drama of the Commons* (eds E. Ostrom, T. Dietz, N. Dolsak, P.C. Stern, S. Stonich, E.U. Weber). National Academy Press, Washington, DC, 233–257.

Ruggiero, R.G. and Fay, J.M. (1994) Utilization of termitarium soils by elephants and its ecological implications. *African Journal of Ecology*, **32**, 222–232.

Sanderson, E.W., Redford, K.H., Vedder, A., Coppolillo, P.G. and Ward, S.E. (2002) A conceptual model for conservation planning based on landscape species requirements. *Landscape and Urban Planning*, **58**, 41–56.

Sist, P. and Ferreira, F.N. (2007) Sustainability of reduced-impact logging in the eastern Amazon. *Forest Ecology & Management*, **243**, 199–209.

Smith, R.J., Muir, R.D.J., Walpole, M.J., Balmford, A. and Leader-Williams, N. (2003) Governance and the loss of biodiversity, *Nature*, **426**, 67–80.

Solly, H. (2007) Cameroon: From free gift to valued commodity – the bushmeat commodity chain around the Dja Reserve, in *Bushmeat and Livelihoods, Wildlife Management and Poverty Reduction* (eds G. Davies and D. Brown). Blackwell Publishing, Oxford, 61–72.

Soule, M.E. and Sanjayan, M.A. (1998) Ecology conservation targets: Do they help? *Science*, **279**, 2060–2061.

Soule, M.E. and Terborgh, J. (1999) Conserving nature at regional and continental scales – a scientific program for North America. *Bioscience*, **49**, 809–817.

Stokes, E.J. (2007) Ecological monitoring program Ndoki-Likouala landscape 2006–2007: Summary of results. WCS, Republic of Congo.

Stokes, E.J., Strindberg, S., Bakabana, P.C., Elkan, P.W., Iyenguet, F.C., Madzoke, B., Malanda, G.A.F., Mowawa, B.S., Moukoumbou, C., Ouakabadio, F.K., Rainey, H.J. (2010) Monitoring great ape and elephant abundance at large spatial scales, measuring effectiveness of a conservation landscape. *PLoS ONE*, **5**, e10294.

Terborgh, J. Nunez-Iturri, G., Pitman, N.C.A., Valverde, F.H.C., Alvarez, P., Swamy, V., Pringle, E.G. and Paine, C.E.T. (2008) Tree recruitment in an empty forest. *Ecology*, **89**, 1757–1768.

Thomas, C.D., Cameron, A., Green, R.E., Bakkenes, M., Beaumont, L.J., Collingham, Y.C., Barend, F., Erasmus, N., Ferreira de Siqueira, M., Grainger, A., Hannah, L. Hughes, L., Huntley, B., van Jaarsveld, A.S., Midgley, G.F., Miles, L., Ortega-Huerta, M.A., Peterson, A.T., Phillips, O.L. and Williams, S.E. (2004) Extinction risk from climate change. *Nature*, **427**, 145–148.

Thorbjarnarson, J.B. and Eaton, M.J. (2004) Preliminary examination of crocodile bushmeat issues in the Republic of Congo and Gabon. Presented at *The Crocodiles, Proceedings of the 17th Working Meeting of the IUCN–SSC Crocodile Specialist Group*, Gland, Switzerland.

Turkalo, A. and Fay, J.M. (1995) Studying elephants by direct observations, preliminary results from the Dzanga clearing, Central African Republic. *Pachyderm*, **20**, 45–54.

Tutin, C.E.G. and Fernandez, M. (1984) Nationwide census of gorilla (Gorilla gorilla) and chimpanzee (Pan T. troglodytes) populations in Gabon. *American Journal of Primatology*, **6**, 313–336.

Tuxill, J. and Nabhan, P. (2001) *People, Plants and Protected Areas: A Guide to In Situ Management*. Earthscan Publications Ltd, London.

Vautravers, E. (2008) La pêche dans les concessions CIB et la commercialisation du poisson de rivière comme alternative à la chasse et complément à l'importation de protéines domestiques pour la société forestière. DLH, République du Congo.

Western, D. and Wright, R.M. (eds) (1994) *Natural Connections. Perspectives in Community-based Conservation*. Island Press, Washington, DC.

White, L.J.T. (1994) The effects of commercial mechanized selective logging on a transect in lowland rainforest in the Lope Reserve, Gabon. *Journal of Tropical Ecology*, **10**, 313–322.

White, L.J.T. and Edwards, A. (2000) *Conservation Research in the African Rain Forests: A Technical Handbook*. WCS, New York.

Wilkie, D.S. and Carpenter, J.F. (1999) Bushmeat hunting in the Congo Basin, an assessment of impacts and options for mitigation. *Biodiversity & Conservation*, **8**, 927–955.

Wilkie, D.S. and Laporte, N. (2001) Forest area and deforestation in central Africa, current knowledge and future directions, in *African Rain Forest Ecology and Conservation* (eds W. Weber, L.J.T. White, A. Vedder, N. Naughton-Treves). Yale University Press, New Haven, Connecticut, 119–138.

Wilkie, D.S., Shaw, E., Rotberg, F., Morelli, G. and Auzel, P. (2000) Roads, development, and conservation in the Congo basin. *Conservation Biology*, **14**, 1614–1622.

Wilkie, D.S., Sidle, J.G., Boundzana, G.C., Blake, S. and Auzel, P. (2001) Defaunation, not deforestation, commercial logging and market hunting in northern Congo, in *The Cutting Edge, Conserving Wildlife in Logged Tropical Forest* (eds R.A. Fimbel, A. Grajal and J.G. Robinson). New York, NY, Columbia University Press. 375–399.

Wilkie, D.S., Starkey, M., Abernethy, K., Effa, E.N., Telfer, P. and Godoy, R. (2005) Role of prices and wealth in consumer demand for bushmeat in Gabon, central Africa. *Conservation Biology*, **19**, 268–274.

Wilshusen, P.R., Brechin, S.R., Fortwangler, C.L. and West, P.C. (2003) Conservation and development at the turn of the twenty-first century, in *Contested Nature: Promoting International Biodiversity with Social Justice in the Twenty-first Century* (eds S.R. Brechin, P.R. Wilshusen, C.L. Fortwangler and P.C. West). State University of New York, Albany, NY, 1–22.

Wood, S.N. (2006) *Generalized Additive Models*. Chapman and Hall, Boca Raton, Florida.

World Bank Group (2010) http://go.worldbank.org/ZP8KW3GA90.

World Rainforest Movement (2003) Congo, Republic: Apes suffer from marriage between loggers and conservationists. http://www.wrm.org.uy/bulletin/73/Congo.html. (Accessed October 2010.)

Wright, S.J., Stoner, K.E., Beckman, N., Corlett, R.T., Dirzo, R., Muller-Landau, H.C., Nunez-Iturri, G., Peres, C.A. and Wang, B.C. (2007) The plight of large animals in tropical forests and the consequences for plant regeneration. *Biotropica*, **39**, 289–291.

Zangato, M.E. (1999) *African Archaeology*. BAR, Cambridge, UK.

Index

Tropical Forest Conservation and Industry Partnership: An Experience from the Congo Basin, First Edition.
Edited by Connie J. Clark and John R. Poulsen.
© 2012 Wildlife Conservation Society. Published 2012 by John Wiley & Sons, Ltd.